Another Chance

II

Claire McClennan

With Drawings by

Linda Thurlow

HRB

HOLROCKS BOOKS

First Published in 1999 by HolRocks Books
PO Box 107
Paignton
Devon
TQ4 7YR

ANOTHER CHANCE II

A CIP catalogue record for this book
is available from the British Library

ISBN 0 - 9526944 - 2 - 5

Printed and Bound in the UK by
Lazarus Press
Unit 7 Caddsdown Business Park
Bideford
Devon
EX39 3DX

"As the old man walked the beach at dawn,
he noticed a young man ahead of him picking
up starfish and flinging them into the sea.
He asked why he was doing this, and the youth
replied that the stranded starfish would die if left
on the beach until morning.

'But the beach goes for miles and there are
millions of starfish,' countered the old man.
'How can your efforts make any difference?'
The young man looked at the starfish in his hand
and threw it to safety in the waves.
'It makes a difference to this one,' he said."

Wildlife Aid

CONTENTS

CONTENTS

RENA & DUKE

THE LIONESS & THE LION

GREEK ANIMAL RESCUE's challenge to help a magnificent Lion and Lioness escape life in a cramped cage, behind metal bars.

Rena lived in appalling conditions on a farm on the Greek island of Kos. Her so-called owner, who already had a collection of other wild animals at his farm, many of whom subsequently died of neglect, had bought her as a cub, illegally, from a gypsies travelling circus.

Vesna Jones, founder of British based Greek Animal Rescue (GAR) came across Rena in July '95, during her yearly visit to the shelters that GAR support in Greece.

Rena's home, where she had lived for almost ten years, was a cage measuring 20 square metres. This enabled her to take only two steps in any one direction. She had no bedding and a dilapidated and smelly caravan, full of old bones and chicken feathers proved to be her sleeping quarters. There was no water and no food in sight. No trees, no grass, just concrete and metal bars.

'Her ears were covered with ticks, she was bored, underweight and lethargic,' says Vesna. 'Her right paw looked raw and both her front paws appeared badly deformed, causing her to limp. My heart went out to her.'

On her return to London, Vesna telephoned Malcolm Dudding, owner of the Big Cat Sanctuary in Kent. He agreed to accept Rena, providing GAR could raise the funds needed for the building of her enclosure – a staggering £30,000, plus the cost of transporting Rena from Greece.

Raising such a huge amount of money for a small charity like GAR, was a daunting task. But *nothing* was going to stop Vesna and so she set to work publicising Rena's plight. Rena appeal posters were circulated, thousands of leaflets were printed and letters were written to the press. But further problems were in store for GAR.

RENA : *Her home for ten years – a 20m² cage.*

Because Rena was being kept illegally, proceedings had to be taken against the owner to confiscate her. With the help of Angela Fleming, an active animal welfarist based in Athens, the injunction was finally obtained in March '97, after a fourteen month battle with legal proceedings and endless bureaucratic red tape.

But the battle to rescue Rena was far from over. The owner and his mob of supporters decided to take the law into their own hands. When Vesna and her team arrived at the farm, with the injunction to confiscate Rena and a truck bearing the crate for her transportation, a threatening, screaming mob of a hundred people surrounded their car and attempted to turn it over.

'Perhaps the most chilling moment for me,' remembers Vesna, 'was when a mean looking thug stopped infront of the car, placed his hands on the bonnet and leant towards the windscreen, all the time staring at my face intently – I was a marked woman! The crate ready for Rena was subsequently doused with petrol and set alight. It was total anarchy!'

Without adequate Police protection, Vesna and her team were lucky to escape unhurt. They had to leave Rena behind.

It was to take another five months before Rena was finally rescued; five months of letter writing, and appeals to MPs and MEPs to support the campaign.

'THE DAY WE HAD BEEN WAITING FOR, FINALLY ARRIVED!' says Vesna. 'Our lengthy campaign to rescue Rena came to an end on 18 August '97 when she was removed from her cage on the farm on Kos. After a twelve hour ferry crossing to Piraeus, she was flown to Heathrow.'

'Words cannot express how I felt when the van containing Rena pulled into the drive at the Big Cat Sanctuary in Kent. Our efforts had finally paid off. It was a dream come true!'

'Here, Rena was to find an enclosure of hundreds of square metres of grass and shrubs, in the midst of acres of fields and trees, with wooden platforms, hilly mounds and tree trunks and fresh straw bedding.'

'Rena did not hesitate to emerge from her sleeping quarters as she began to explore her new home for the first time. She walked around the enclosure slowly, smelling the grass, the trees and the shrubs.'

'Rena gazed across the field, stretching her neck as if to see better. She was alert yet at the same time quite relaxed and looking very much at home. At one point, as she was lying down with her two front legs stretched out, her stumpy paws gently started to knead the grass. She began to playfully roll on the soft turf. It was so moving to watch her – tears welled up in my eyes.'

'The battle to rescue Rena had been long and hard, fraught with difficulties and endless red tape, not to mention frustrations. It wasn't the stubborness which kept us going, but the knowledge that, should we have failed, Rena would have been doomed to spend the rest of her life in that cage, her deformed paws would never have received treatment and she would have died never knowing what it felt like to walk on soft green grass.'

RENA : *Playing, chewing, rolling and happy with her catnip cushion.*

RENA : *Responding to Vesna's gentle banter*

RENA : *Surveys her new home.*

Duke – The Sad Lion.

Duke was a very sad, underweight, lethargic Lion. He came from an Italian travelling circus and his home for three years had been at the Philadelphia Zoo in Athens – a twenty-five square metre concrete pit with drab concrete walls.

All day, every day, Duke drooped his head and gazed through the metal bars.

DUKE : *So miserable in his concrete pit.*

His one and only friend was Tommy. Tommy, the kitten, wandered one day into his cage and a most unusual friendship developed. Duke tolerated him simply because he was bored and lonely in his barren cell.

When GAR heard of Duke's plight, they did not hesitate to adopt him. Rena's enclosure at the Sanctuary was built to accommodate two big cats and GAR felt it would be good for Duke and Rena, being the same age, to have each other for company.

On the day Duke was transported to England, his friend Tommy kitten was not forgotten. He was collected by a GAR representative in Athens, treated for cat 'flu and is now a healthy, happy cat, the memory of his big brother Duke long forgotten.

DUKE & TOMMY : *Captive Friends.*

'Duke now shares the enclosure at the Sanctuary with Rena,' says Vesna. 'Watching Duke strutting around the enclosure, head held high with all his senses alive, is a sight to behold. He looks truly magnificent! His coat is shiny, his mane immaculate, bushy and silky, the pockmarks on his back and sides that covered his body have completely disappeared but the most noticeable change is the contented look in his eyes! Not so long ago those eyes were dead, expressionless, filled with boredom and loneliness. His days in the zoo cage had been spent mostly sleeping – there was nothing for him to do or look at. He was a very sad lion indeed. We have given him the opportunity of a much happier life.'

'By rescuing Duke and Rena, GAR has undertaken the responsibility to provide them with the best possible care for the rest of their lives. Raising several thousand pounds each year for their keep will not be an easy task for our small charity but we will do our best.'

'GAR was founded primarily to help injured and abandoned stray dogs and cats but when we heard about Rena and Duke, we couldn't turn our backs on their plight.'

'They say: "WHERE THERE'S A WILL, THERE'S A WAY". We have proved this to be true. Rena and Duke have become very good friends, often lying side by side and nuzzling against each other. Rena enjoys chasing Duke around biting his tail and running off. He replies with a roar and proceeds to chase her. Rena is now a very contented Lioness; Duke walks tall, a very proud beast.'

RENA & DUKE : *Carefree and content.*

DUKE : *A proud, majestic beast.*

ROWAN

The tiny Roe Deer Fawn, given a helping start to life by WINSLADE WILDLIFE SANCTUARY.

Nestling amongst remote farmland, surrounded by twisting country lanes, is a busy refuge for many injured rabbits, ducks, swans, guillemots and hedgehogs. Leafy trees, shrubs and bushes surround the enclosures, while an intensive care unit is furnished with twigs to resemble the patients' natural habitat.

This is Winslade Wildlife Sanctuary at Buckland Brewer, and in this tranquil setting a little Roe Deer Fawn found some friends.

Whilst out walking, a lady had come across the tiny fawn. He was all alone and she sat with him for two and a half hours waiting for his mother to return! But with no sign of the mother, the lady decided to bundle him up in her coat and rushed him to the nearest vet in Barnstaple. Within a matter of hours, the little fawn found himself entering the gate at the Winslade Sanctuary.

'We named him Rowan,' says Tracey Berridge, Deputy Manager, 'due not only to the fact he was a Roe Deer but because a Rowan tree lives a very long life and we hoped this little fawn would indeed have that very long life.'

'Rowan almost died a couple of times but managed to bond with Judy Greisbach, the Sanctuary's founder. She became his temporary mum. She stayed with him constantly for two days and wouldn't allow any photos or cameras near him until he was at least five-weeks-old. She saw him through the crucial period of his life, offering friendship and peace – if there is too much noise, fawns will die from stress-related disorders.'

'Rowan stayed with us at the sanctuary until he was eight-weeks-old and everybody cried when he left.'

But Rowan was ready to meet other fawns and off he went to West Hatch in Taunton – the RSPCA Wildlife Unit – to begin the second stage of his rehabilitation into the wild. Within a few months he was released successfully into the Roe Deer herd.

This was the result Judy had hoped for. Her priority, in all cases, is to return the creatures to their natural environment.

Judy has been running her wildlife sanctuary in North Devon for almost twelve years and receives up to one thousand wildlife enquiries every year. Helped by volunteers from the probation service, from those who want work experience in animal care and from the Prince's Trust – whose volunteer team recently built the sanctuary a new aviary – her days are spent caring for sick and injured wildlife and educating people about wild animals.

Winslade is kept going entirely through voluntary donations and fund raising. Supporters wearing wacky green and yellow 'duck' hats go out and about in North Devon with their collecting boxes, accompanied by Jay The Clown in his sunflower suit, brandishing his yellow and red umbrella and feather duster. Everyone clowns around – all to help the wildlife!

ROWAN : *The Baby Fawn.*

©NAVS

A very, very lucky pup.
One of six, rescued from a breeder who
supplies vivisection laboratories.

POPPY, JONESEY, PASCA, NAT, CLEO & ERIC

THE BEAGLES

In an undercover operation, NAVS came to the rescue of these Six Beagle Pups, bred by a laboratory supplier and destined to be used in horrendous experiments.

Poppy, Jonesey, Pasca, Nat, Cleo and Eric are six *exceedingly* lucky beagle pups. They are just six of the nine thousand dogs that are used in animal laboratories in the UK each year. The majority of laboratory experiments tend to be performed on beagles reared in intensive farming conditions by specialist suppliers and the UK has become a major international centre of trade for these dogs.

The National Anti-Vivisection Society recently set out to expose this sordid trade. They had discovered that a certain specialist supplier in Cambridgeshire had become Europe's largest dealer of laboratory beagles, sending 2,500 puppies to their fate in laboratories all over Europe, the Middle East and the Far East. NAVS evoked a plan to investigate the activities of the beagle supplier and, at the same time, save at least one consignment of pups from their terrible fate.

'Firstly,' says Jan Creamer, Director of NAVS, 'we placed a worker in a laboratory to investigate the day to day life for laboratory animals. Here, dogs are used to test everything from household products to agriculture herbicides. In one of the tests we secretly filmed, the beagle pups were force-fed a herbicide which had already been animal tested and passed as safe under UK legislation and which has been for sale for twenty years. The beagle pups for these tests are supplied by the specialist supplier in Cambridgeshire. So NAVS working together with The News of the World, approached the supplier and set up a front. "Operation Release" had begun!

'The supplier took the bait,' says Jan. 'Believing we were buyers and he was on the brink of securing a regular order of beagles for an Italian Laboratory, he took me, together with a reporter from The News of the World, to see his dog units.'

'We saw hundreds of dogs housed in metal pens, having no contact with the outside world. As he showed us around, a mother beagle was giving birth in a hard concrete pen. He talked readily about how the vivisection laboratories demand pups between three and six months old. He claimed he produced almost a million animals for experiments every year – dogs, guinea pigs, rabbits, rats and mice. (This amounts to two animals for research every minute.)'

'As the supplier showed us around, six little pups, between four and five months old, were being loaded into the van by the driver – in reality, another member of our team. Instead of making their first step on the road to death in a foreign laboratory as the supplier believed, these six *very, very* lucky puppies were actually making their first steps to freedom.'

'At their overnight stop, the six pups cautiously and nervously explored a whole new world of soft beds, toys and kind words. The next day it was time to see grass, the sun and the sky for the first time in their short lives. After fifteen minutes of cautious exploration, the joy of it all took over as they raced around like any other puppies. Laboratories often claim they cannot re-home animals, even if experiments are not terminal, because the animals have been bred for life in a laboratory and they would not be happy in an ordinary family home. It took Poppy, Jonesey, Pasca, Nat, Cleo and Eric less than twenty-four hours to revert to type.'

'It is only with the generous donations from the supporters of NAVS,' says Jan, 'that we are able to help these animals and conduct our educational and publicity campaigns.'

Through the work of the Special Investigation Department, this bundle of beagles, instead of living through a few months of pain and suffering leading to death, have been given another chance in happy, new family homes.

PASCA : *Peeps with a friend from his travelling crate during the rescue.*

©NAVS

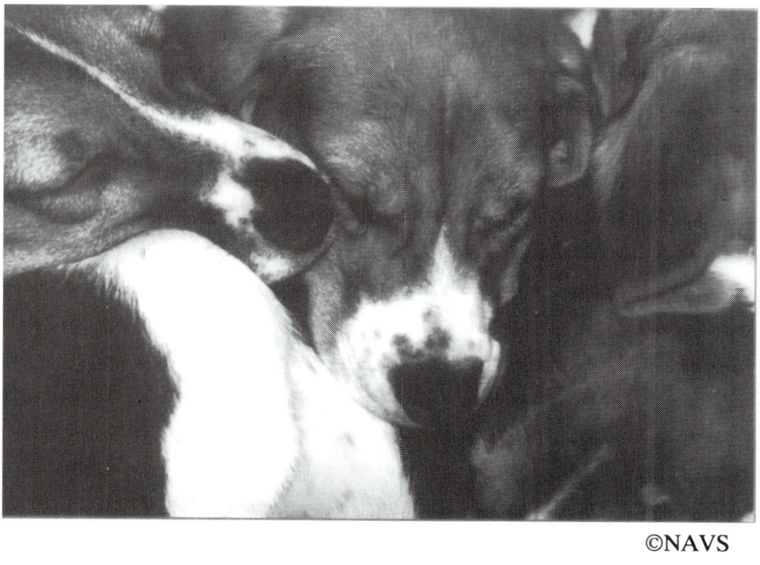

©NAVS

The snuggle of Beagles rescued by NAVS.

SOBHRACH & ROSIE

THE COW & HER CALF

Rescued by an elderly lady who did her utmost to keep them together. When all her attempts failed, the HILLSIDE ANIMAL SANCTUARY came to her aid and saved her friends from the farming industry.

For many years, Wendy Valentine's name has been synonymous with the care and concern for horses. She has worked tirelessly on both sides of the Atlantic on behalf of these beautiful, much abused animals.

Wendy has now turned her relentless energy towards making the public aware of the suffering involved in intensive factory farming. In 1995 she founded Hillside Animal Sanctuary to help all animals in need, especially those hidden from public gaze involved in food production.

When she heard of the plight of Sobhrach and Rosie, she offered a place at her sanctuary without a second thought.

'A couple of years ago,' says Wendy, 'we were asked to take in a rescued cow and her calf by an elderly lady who was finding difficulty in looking after them due to her failing sight and a lack of proper facilities to keep them as she would have liked. She was terrified that if anything happened to her, they would end up back in the farming industry from where she had already saved them.'

Sobhrach (Gaelic for "Primrose") and Rosie lived on the Isle of Skye, a two day journey from the Hillside Sanctuary in Norfolk, but regardless of this, arrangements were made. Moving day came, but when the transport arrived to collect the animals, the wonderfully caring owner could not bear to see her two friends leave. She turned the lorry away, empty.

She had a love for this cow and calf that was so great that she was determined that *somehow* she would struggle on to keep them with her. She took such desperate steps for them. She decided to give up her home and she moved into a croft with more land on a nearby island – The Island of Lewis. Wendy assured her that should the time arise again, Sobhrach and Rosie would always be welcome at Hillside.

Several months went by, when Wendy received a further letter from the now very worried elderly lady. It said :

'The move here has turned out to be a disaster. When, the last time I tried, I simply could not let Sobhrach and Rosie go, I gave up my home and bought this four-acre croft, hoping to be able to keep them with me. The building society did not tell me, until winter was upon us, that the little house had had such a bad survey that they could not give me the mortgage I required – which I had intended to spend on shelters for my animals. So I adapted a large shed for Sobhrach and Rosie and stored their hay and feed in my sitting room!'

'I also knocked up a sheep shelter, only to find that both shed and shelter were severely damaged on more than one occasion by storm-force gales. I did not realise that the land had been reclaimed from the ocean and that the ocean was attempting to take it back!!'

'I woke one morning last November to find my sheep up to their chests in water and Sobhrach up to her udders. I brought them up to the front garden, padlocked the gates and built a style for callers, then set about getting the croft drained.'

'Sobhrach and Rosie have been really good and made the best of it but have spent most of the winter inside the shed. They have never been tied up and of course, I have given them lots of straw to lie on and hay to eat (and molasses, bruised oats and vitamins) but they have not liked going outside and standing knee-deep in thick mud. Occasionally they have tried to pick their way round the edges but it is miserable and really not good for them. Apart from this, the weather is too wild up here for cattle to survive (there are a few but several of these died this year) apart from the indigenous Highland breeds. The vet is not very accessible and I dread to think what would happen to them up here if anything happened to me.'

The letter continued : *'As you know, I am adamant that they don't go into the farming industry – so the alternative would be to have them shot. You are the only person, Wendy, I would trust with them as I know you love and care about animals as I do. I think, for their sakes, the time has now come where I have no choice but to send them down to you. I have spent the winter trying to find some way of keeping them, but without success. All the crofts here get water-logged (but mine seems to be one of the very worst).'*

'The thought of sending them away is making me quite ill and it will break my heart to let them go – but for their sakes I know I must do so. How I shall survive without them I really don't know – as every morning they lick me and bury their heads in my arms. There is such a bond between us. They know I love them and they must trust me – so it is quite dreadful for me.'

Hillside Sanctuary immediately opened it's doors. In May '97, Sobhrach and Rosie began their long journey, from their croft home on the Isle of Lewis, to Norfolk. Resting overnight, they arrived safely the following day at Hillside. 'Half of me has left with them,' said their owner.

'Here they will live out their days together in peace,' says Wendy. 'I find it amazing how one kind lady can go to such great lengths to give her much loved cows all she possibly could while so many other people can only exploit these farm animals. We at Hillside, believe animals have an inherent right to live and die with dignity and in peace.'

The story has a very happy ending for Sobhrach and Rosie. They joyfully spend their days grazing in the Sanctuary meadows with a wonderful assortment of rescued animals to keep them company. Their owner too now has peace of mind. She hopes to make the long journey down to Norfolk several times a year and no doubt will receive a knowing, loving welcome from her grateful best friends.

© Hillside Animal Sanctuary

ROSIE & HER MUM SOBHRACH :
Content to remain together to graze the green grass at Hillside Animal Sanctuary.

BARBARA
THE ASIAN ELEPHANT

Captured from the wild to tolerate a life performing in captivity. Helped by Carol Buckley of THE ELEPHANT SANCTUARY, Tennessee and Sharon Callahan, she was restored to health and returned to green pastures.

Sharon Callahan is an Animal Communicator and a leading pioneer in the use of flower essences in the treatment of animals. Her experience in these fields has gained the respect of individuals, animal welfare organisations and flower essence professionals throughout the world. She has worked amongst animals and with the people who love them, for more than thirty years.

'When I was a child,' she says, 'I was filled with animals. From the earliest time, I can remember I was shocked by the ways in which animals were treated by human beings. Animals were always with me, one way or another, although relegated to hobby status. It wasn't until my forties, that the animals returned to my life full force. It was a long, serious illness that opened the door for their return.'

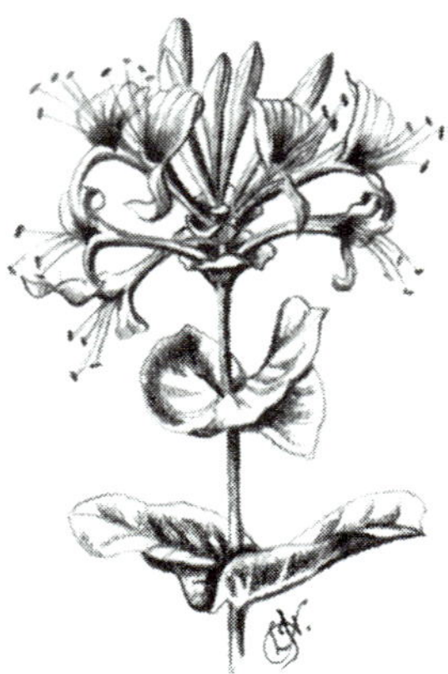

After a near-death experience, Sharon restructured her life to include more time for communion with nature.

'I would sit outside with my two cats, or walk in the woods and I would hear the animals, plants, flowers and trees talking with each other. This enhanced my ability to communicate with animals telepathically and to connect with their souls. I combined my love of flowers with my ability to communicate with the animals to create a system of healing.'

Sharon created Anaflora – the original flower essence formulas.

She designed them specifically for treating both domestic and captive wild animals and collected the wild flowers herself from around Mt. Shusta, her home in California.

'The flower essence,' she says, 'embodies the specific energy pattern of the flower from which it comes. This unique life force pattern of each flower, when taken as an essence, awakens qualities that have the power to override negative energies present in the individual who takes them. Animals that come from abusive or neglectful backgrounds often require assistance to clear negative patterning.'

'When I learnt about the healing properties of herbs and flowers – animals in the wild will seek these out – it seemed natural to offer the essences to my animal friends too.'

'Barbara is one of many animals who has touched my heart, to whom I have offered help and love.'

Barbara, a female Asian Elephant, was born in 1972 in the wilds of South-East Asia. When she was barely two-years-old, she was singled out because of her age, beauty and disposition and was captured by mahouts – Asian elephant trainers. Riding on trained elephants, the mahouts entered her herd, dropped ropes around her neck and legs and dragged her from her family and jungle home.

'Barbara,' continues Sharon, 'was taken to a logging camp and tied to a tree for several weeks while she learned to tolerate life in captivity. She was then purchased by a circus and loaded into a wooden crate and sent to America.'

'For several months, Barbara and seven other elephants were taught a variety of tricks – standing on their hind legs, holding one another's tail and lying on their sides. Due to their great intelligence, the tricks were not difficult for the elephants but the work was unnatural and very stressful. They would spend up to eighteen hours chained in a stall, being fed only once a day.'

'Year after year, Barbara and her new elephant family performed in travelling circuses throughout the USA.'

At fourteen years of age, Barbara and the other elephants were retired to a farm in Florida for breeding. During this time, Barbara began to lose weight. Many tests were performed but all came up negative and no answer could be found for her loss of weight. For many years she suffered with this weight condition, losing a total of 2,000 lbs.

At the breeding farm, several baby elephants were born during this time – but none to Barbara.

The elephant family then moved on to a new breeding facility. Looking totally emaciated, Barbara was separated from her family and put in solitary confinement. This enforced separation caused her to enter a deep depression.

But finally, at the age of twenty-four, Barbara's luck changed. She was permanently re-located to The Elephant Sanctuary in Hohenwald, Tennessee. This was the only natural habitat refuge for Asian elephants and Barbara was to join the other resident elephants – Jenny and Tarra. Here she could roam green pastures and bathe in spring fed streams.

With her own savings and donations from friends, Carol Buckley had set up this Elephant Sanctuary as a natural habitat refuge, where sick, old and needy elephants could once more walk the earth in dignity. Her growing interest in elephant welfare came when, knowing elephants from zoos and circuses, she realised she most enjoyed elephants when she could take them out in the woods and just let them be. She had always had just one wish : 'If I could do anything, I'd just be with elephants and just let them *be* elephants.'

'As an endangered species,' says Carol, 'Asian elephants are entitled to protection. As highly intelligent animals, who have spent their lives serving and entertaining mankind, they are especially deserving of our compassion and respect.'

So, it was Carol who offered to take in an emaciated and lethargic Barbara. She became her handler and her friend. And because of her willingness to try anything that might help the elephants, Carol agreed for Sharon to treat Barbara with flower essences to help restore her health and character.

Sharon began a telepathic communication with Barbara and gave her a variety of remedies containing Star of Bethlehem, Morning Glory and Fringed Violet. For an animal the size of Barbara, the essences were prepared in extra large bottles!

BARBARA

BARBARA, JENNY & TARRA: Freely grazing the pastures of the sanctuary.

SHARON CALLAHAN : *'All through my life animals have been my wise friends and spiritual companions. I feel at peace around animals. Animals have been my greatest teachers and my life has been profoundly enriched by their presence. From them I have learned about joy, spontaneity, humour, acceptance, surrender, devotion, fidelity, death and resurrection. I have begun to realise that angels have whiskers and wet noses!'*

Sharon remembers : 'After her first treatment, Barbara would open her mouth and stick out her tongue when I approached with the essence bottle! She soon began to relax a little and she soon trumpeted, perhaps for the first time in her entire life! We watched her bathing and playing in a mud hole like a bird at a bird-bath. When offered words of encouragement, she responded with more playful gestures. Barbara had been tentative about affection when she had arrived at the sanctuary but as the days went by she began to "lean into" Carol and become more playful.'

Soon Carol was reporting back to Sharon that Barbara was "effervescent" and "has a twinkle in her eye almost all the time"! She developed a playful relationship with one of the sanctuary dogs and when she settled down, she began to talk in the way elephants do when contented and happy. By now, Barbara was lying down beside elephant Jenny, in true elephant comradery, with trunks entwined. With elephant Tarra, Barbara soon found she had a sound friendship.

Sharon continued to administer various flower essences and talk telepathically to Barbara. Progress was excellent, she continued to gain weight and Barbara would soon allow herself to be rubbed all over – even the scar around her neck, that was untouchable before.

'Barbara is coming into a sense of herself now,' says Sharon. 'To this day, she wears a one-inch-wide deep ring around her neck – a testament to her struggle as she was dragged from her family by the ropes of the mahouts. But all three elephants have become friendly and well adjusted.'

They are now looking forward to the introduction of two new elephants and the sanctuary hopes to support an additional ten in the future. Carol's dream is that her sanctuary should become a prototype for others to follow.

MINNIE
THE CAT

One of LIFE FOR CATS very special rescues.

© Life For Cats

The Life For Cats charity was established to provide shelter for unwanted, homeless and abused cats. Minnie came into all three categories.

'Minnie was found wandering the streets, bald from the waist down,' says Peri Ann Sinclair. 'When we took her into our shelter, she was full of fear, trembling and petrified of any human contact.'

'We shall never really know what Minnie truly suffered but after months of care and love, Minnie is healthy again. Minnie has become very special to us.'

SARAH
THE KITTEN

The tiny traveller discovered by
SOUTH DEVON ANIMAL SHELTER.

© SDAS

A very soft mewing sound could be heard coming from the engine of a Ford Sierra. Sarah, a very frightened little five-week-old kitten was found trembling under the bonnet, having travelled along for a good ten miles.

Quite unharmed, she was safely rescued by South Devon Animal Shelter.

'How she came to be there or where she came from, is a mystery known only to her!'

GEORGE
THE FOX CUB

When given a chance by WILDLIFE AID, he showed a determination to survive beyond measure.

Wildlife Aid is a leading Wildlife charity, dedicated to the rescue, care and rehabilitation of sick, injured and orphaned British wildlife.

Their voluntary helpers find themselves up all night feeding baby bats, every hour-and-a-half right around the clock; then there are the two-hourly feeds of orphaned fox cubs, baby owls and as often as not, midnight rescues. Rescuing trapped herons from netting on garden ponds is a skill in itself. In this, George Holewski, at a macho 6ft. 2ins., has become something of a hero as he wades through very deep, very cold water.

Malcolm Thompson is Wildlife Aid's main large animal rescuer – the man *always* on the spot. He has followed fully grown roe stags, savaged by dogs, through rivers and turned into Superman when in hot pursuit of a Canadian goose! He took off across the roof tops of British Home Stores in Kingston in full pursuit of an adult fox and spends many an hour dripping in dirty pond water! Four words ring constantly in his ears: 'Malcolm.... Are you busy?' But he is the first to admit, life as a rescuer can be very rewarding.

One recent rescue goes to show what Susan Blissett of Wildlife Aid, has been telling the public at every possible opportunity.

'A wild animal lying in the road is not *necessarily* dead. It is well worth checking; it may only be injured or unconscious. Quite often it is the second or third car that comes along that delivers the fatal blow. Our little cub George, as is so often the case with foxes, once given a chance, showed a determination to survive that was beyond measure.'

Early one Sunday morning, a couple were driving through the Surrey countryside, when they spotted a fox cub lying on the side of the road. They stopped their car and got out to move the supposed dead creature off the road and into the kerb.

© Wildlife Aid

Feeding a Hedge Hopper -
Wildlife Aid's baby Little Owl.

© Wildlife Aid

A pair of Baby Owls.

‘At this point,’ says Susan, ‘the couple noticed that the little cub was, in fact, still breathing. They knew our 24-hour helpline number and called us. I was on duty in the hospital that day and advised them to put an upturned dustbin over the cub to protect it. I then dispatched one of our volunteers.’

‘This little cub was about ten-weeks-old and had a very severe wound to the top of his head. He was unconscious so I tenderly felt all around the scalp to make sure that it wasn’t fractured. Fortunately, it wasn’t. So I set to and cleaned up the wound to ensure no infection would set in.’

George, as he had been named, was put in the Intensive Care Unit, under a heat lamp and left to sleep. About four hours later there were signs of movement and the little cub, with Susan supporting his head, managed to lap at a warm bowl of glucose water albeit in an uncoordinated way.

‘By the end of the first day, George was standing rather wobbly and had managed a little food. Progress indeed!’

Three days later George had moved up to one of the outside pens. Within a week his head wound was stitched and by the time he was released back into the wild (with three other rescued cubs that had been placed in a happy family pod) there was not a mark to be seen.

‘All our rescued fox cubs undergo a ‘soft’ style release. We build large holding pens in ‘friendly’ non-hunting areas and the cubs are moved there in the Autumn. After a month, we open the door and the cubs are free to ‘escape’ in their own time as food is made available in the pens for as long as necessary. The cubs disperse very successfully and it gives them all the time they need to adapt to the rigours of living in the wild with quiet assistance from us.’

George went off on his way, happily back into the wild.

©Wildlife Aid

GEORGE : *The Little Fox Cub.*

'He left bright eyed and very bushy tailed -
as all self-respecting foxes should!
We certainly hope he learnt his lesson about cars!'

PLYMMIE - THE PUPPY
EROL & GULER - THE TWIN BEAR CUBS

How WSPA came to the rescue of an enchanting puppy, abandoned beneath the erupting volcano on the Caribbean Island of Montserrat.
They were on hand too for tiny twin Bear Cubs, destined to be sold into slavery as dancing bears in Turkey.

The Caribbean Island of Montserrat used to be, quite simply, paradise on earth. But since July '95, this Caribbean paradise has suffered constant eruptions of rock and ash from The Soufriere Hills Volcano. The island has been transformed into a grey wasteland. Some of the islanders chose to stay on but thousands of people have been evacuated and thousands have lost their homes fleeing to safety.

The World Society for the Protection of Animals (WSPA) had one immediate concern. In the ensuing panic, the animal victims of this disaster – the livestock, the pets and the strays – were being left behind without food or water. So they organized for huge quantities of pet food and thousands of dollars of veterinary medicines to be shipped out. Visits were made to pinpoint locations of abandoned animals and temporary shelters were specially constructed.

Gerardo Huertas of WSPA said : 'Lost dogs and cats have been the forgotten casualties of this disaster, with many animals left abandoned or lost in the deserted ash-strewn landscape. Until WSPA came along, these animals were in danger of starving to death or being killed by the red-hot lava flows.'

The former capital Plymouth, had been totally destroyed by volcanic eruptions but it came to WSPA's attention that a lost puppy had been seen roaming in the deserted ruins of the town.

Jonathan Owen, part of WSPA's Animal Rescue Team said : 'We knew we *had* to try to rescue the town's last surviving animal. We headed off for Montserrat.'

©WSPA

The Soufriere Hills Volcano looms over the ruins of Plymouth, the former capital of Montserrat.

'As we sped towards the volcano by speedboat, the lush green coastline quickly gave way to a grey, dusty landscape. We felt a strange atmosphere of foreboding. As the dark silhouette of the volcano came into view, we realised the grey clouds were caused by outpourings of ash, like steam from a constantly boiling kettle.'

The rescuers made their way to Plymouth, near the foot of the volcano and well inside the dangerous 'exclusion zone'. Vast amounts of volcanic ash covered the town, over eight-feet deep in places, topped with a thin crust. This had to be delicately negotiated to avoid falling through to the hot ash underneath.

'We were aware of the need to work quickly,' says Jonathan. 'I could feel the heat from several feet below coming up through my boots and in some places the volcanic ash was still too hot to walk on. Our progress was painstakingly slow. The silence weighed heavily in the volcano's shadow only broken by the occasional low rumble from the volcano itself and our shouts and whistles for the lost puppy.'

Volcanic ash continues to fall over the island during the rescue team's visit.

© L.Bopp / WSPA

The rescue team wandered through the charred ruins of the town. The collapsed buildings were strewn with debris from fallen power lines; street girders had been twisted by the volcanic heat; telephone boxes stood half-full of ash with melted handsets. In this ghostly setting the rescuers searched and searched.

'Just as we were wondering if we would *ever* find the puppy,' says Jonathan, 'I suddenly saw a brown dog haring towards me, it's tail wagging furiously! As it raced past, I shouted to the others who arrived breathlessly to see an overjoyed puppy dancing around us in excitement!'

John Walsh, WSPA's Director of Animal Rescue, had come prepared. With patience and experience he very slowly managed to gain the puppy's trust – with the aid of cheese and sardines!

'Hardly daring to breathe,' says Jonathan, 'we watched the puppy edge nearer, until the moment when John snatched her up into his arms. We cheered and applauded! We had saved the puppy from certain death. John cradled her in his arms as she licked his face and wagged her tail furiously. I had never seen a dog so happy to see someone! We chose to name the puppy Plymmie – after the town of Plymouth.'

©WSPA

PLYMMIE : *The only remaining resident of Plymouth, excitedly greets John Walsh, one of her rescuers.*

Mission accomplished, the rescue team set off to return to the safe zone. Within minutes, a tropical storm broke. They stopped for a moment to take a backward glance. Behind them was the awesome sight of the volcano spewing a river of white-hot ash down towards the town. Only minutes before, the entire rescue team had been searching in that area.

They took Plymmie across to the 'safest' part of the island, to a temporary shelter built by WSPA, where she joined over seventy pet dogs and cats, rescued on earlier missions. Many had been given emergency veterinary care for very sore eyes, due to the down-pouring of ash. Each animal was checked, registered for evacuation and given their own registration cards and tags.

'A frenzy of barking greeted us as we began to evacuate the dogs from the shelter,' says Jonathan. 'In their portable kennels, they were taken to Little Bay Harbour and loaded on to a ferry to take them to the nearby island of Antigua, three hours away.'

'From Antigua, they were put on to a special freight 'plane and flown to Miami, Florida. Here they were all to find special, new homes. Interestingly, their arrival from Montserrat has had the effect of tripling the usual number of visitors to Florida's dog shelters, resulting in record numbers of stray and abandoned dogs being found new homes as a result.'

As for Plymmie – he was adopted by John Walsh, his rescuer, and along with Elizabeth, John's other dog, has a wonderful new life living near Boston, Massachusetts.

Says Jonathan : 'Several thousand people still remain with their pets on Montserrat but the volcano continues to grow, sometimes as much as seven feet in a single day. Two-thirds of the island has now become too dangerous for people to live on. But people there have begun to realise that dogs and cats are part of their family and what we are doing is actually good.'

PLYMMIE : *With John and Elizabeth, at her new home, far from the constant eruptions of volcanic rock and ash.*

*A surviving dog receives a much needed meal.
His WSPA rescuer beside him is donned in a mask, worn for protection against the volcanic dust and ash.*

©L.Bopp / WSPA

©WSPA

*All Aboard for Antigua!
The evacuees leave the white hot ash on Montserrat behind them as they set sail for their new lives in caring homes.*

The WSPA charity is not just involved in rescues – much work is done by undercover investigators exposing cruel animal practices.

In many parts of the world, the trade in dancing bears is outlawed under wildlife protection acts but WSPA believes that thousands of bears are still kept in captivity and forced to perform for tourists.

In the forests, the mother bear is stalked by poachers. When she goes to forage for food for her young, the poachers strike. Her four-week-old cubs are snatched, their noses and cheeks pierced with red-hot irons and rope is forced through the holes. Their teeth are removed and the tiny cubs begin a regime which involves massive pain. They are starved into submission and they are led on a rope in circles for hours a day, until they learn not to pull. The cub will raise itself on it's hind legs, to reduce the pressure on it's muzzle and raise it's foot, to avoid being lashed by the trainer. Soon, the tapping of the stick becomes enough to make it lift it's feet in fear. Within two years, the cub has learnt it's role – it is now a dancing bear. Before their ninth birthday, due to little food and cruel treatment, these cubs will be dead. In the wild they would have lived to three times this age.

By chance, WSPA's arrival in Turkey, saved two little bear cubs from this terrible fate.

© WSPA

'This way to Freedom!'

One day, one of these little bears escaped from her home – a busy tea-shop in downtown Istanbul. Owned by a wealthy Turkish businessman, she would spend her days playing with the tourists.

Having made her escape, she was seen wandering alone, lost amongst the traffic in the busy, noisy roads of Istanbul.

When finally the Police and the Turkish Animal Welfare Society managed to catch up with her, she was taken to the safety of WSPA's Turkish Bear Sanctuary.

Back at the tea-shop, her twin brother, an equally charming, tiny bear cub, continued to run around and play in the paved courtyard, much to the tourists' delight.

Jonathan Owen and the WSPA Rescue Team soon found themselves in Istanbul, heading for the busy tea-shop, their aim being to persuade the businessman to release his baby male cub, so that he could be reunited with his twin sister and live in the safety of the Bear Sanctuary.

'The owner claimed that the two cubs had been given to him as a gift by a business partner,' says Jonathan. 'The tragic truth, as I pointed out to him, was that their mother was almost certainly shot and killed to enable a hunter to steal the cubs from her to sell as exotic pets. Such orphan cubs, once they grow too big to handle, are destined to a life chained in captivity or being sold into slavery as dancing bears.'

After many cups of hot sweet Turkish coffee, and several hours of negotiations, Jonathan sought to persuade the sceptical owner that it would be best for the cub if he voluntarily surrendered it into WSPA's care.

'Keeping a bear is illegal under Turkish law,' says Jonathan, 'but we were relieved finally to be able to take the cub without involving the police.'

A growing crowd of onlookers arrived as the baby cub was placed in a travel box in the back of the car, which soon set off on the long journey. Crowds continued to gather around the car throughout the journey, fascinated by the sight of the little bear cub.

'When we arrived at the Bear Rehabilitation Centre, several hours later,' remembers Jonathan, 'my fingers were numbed by the cub who had been suckling them furiously for most of the journey!'

The workers at the Centre cheered the baby cub's arrival and everyone watched as he was released into the outside area where his sister – who had arrived a few days earlier – was waiting.

‘After the initial shock and excitement,’ continues Jonathan, ‘they became inseparable. They rolled around in the long grass, the air punctuated with a series of little grunts as they wrestled and fought. Only the temptation of milk and honey, persuaded the cubs to ‘cease hostilities’ to be bottle-fed, purring loudly, their tiny paws clamped around the bottle.’

‘The cubs quickly explored their new surroundings, pausing to sniff the air or explore a tuft of grass or single flower, before turning their attentions to us. As one scampered over to try to climb up on to my knee, I suddenly felt a paw pulling at my T-shirt from the other cub who had crept up behind.’

‘As nightfall came, we took the two cubs into their pen, where they tussled and rolled around together, before finally falling asleep wrapped around each other – an exhausted bundle of fur, sticky with honey and covered in straw!’

The two little cubs were named ‘Erol’ – after Libearty supporter Erol Erim and ‘Guler’ – in memory of his late wife who loved bears.

After several weeks at the Rehabilitation Centre, Erol and Guler were taken to a special pen within the forested Sanctuary at Karacebey, South-East Turkey. Here, all the adult bears gathered around in fascination, as the twin cubs fought and played with boundless enthusiasm.

Since their arrival, WSPA have rescued a further three cubs, who have now joined the twins. They plan to build a special nursery area, next to the Sanctuary, where the cubs can live and grow.

WSPA have *almost* eradicated the barbaric spectacle of dancing bears in Turkey. Over sixty bears now enjoy a new life at WSPA’s own Sanctuary up in the forested hills. In this vast landscape they have pools to splash in, trees to climb and acres of forest to explore. With continuing support from the public, they plan to develop further Sanctuaries in other countries, to save *all* the suffering, dancing bears. Hopefully, other cubs may be as fortunate as Erol and Guler.

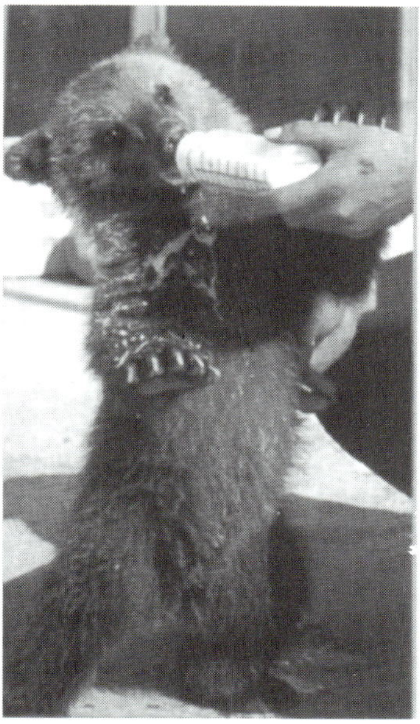

A spot of sustenance after a long journey to freedom.

© WSPA

© J.Hutchison / WSPA

The Twin Terrors – re-united at the Sanctuary!
'As they rolled and played in the long grass, the air was punctuated with a series of little grunts.'

ALICE IN WONDERLAND
THE BROWN HOODED WHITE RAT

Who found refuge in NOAH'S ARK.

Noah's Ark is a foster home for many small, unwanted animals, caring for them until a permanent loving home becomes available. Rabbits and Guinea Pigs, Hamsters and Canaries, Finches and Chinchillas are all found aboard the Ark, cared for by humans who love *all* animals.

One summer, a couple of years ago, a pet shop in the local area was threatened with closure. One hundred mice urgently needed re-homing. Beverley Taylor, the Ark's founder, against her better judgement, opened up the doors and agreed to take the mice in. She was told all the mice had been separated into groups of males and females but this was not true and before long the Ark was over-run with heaps of baby mice! 'Very cute,' thought Beverley, 'but now an even bigger problem with re-homing!' However, with the local media's help, 'The House For A Mouse' campaign began which was a huge success. All the hard work was worthwhile and by autumn, all the mice had scampered off to their new homes!

Recently, The Rat Rescue Centre took up residence within the shelter of the Ark, educating the public on how rats really do not deserve the bad press they receive. As Bev says : 'Although hated and feared by many, the pet rat makes a wonderful pet. Small, odourless and easily trained, they can be the perfect companions.'

'As a rodent, small animal sanctuary, Noah's Ark has re-homed and helped countless rats – well over one hundred in our first year. Not as visible or as cuddly as the traditional dog or cat, few people care about pet rats but rats nevertheless suffer as much neglect and cruelty as any other type of pet animal. The story of Alice is typical of our many rescued rats.'

At her owner's home, Alice spent every spare moment in her cage chewing and chewing. Being inquisitive, she eventually made a hole, large enough to squeeze through – only to encounter

BEV : *With her three little Rats.*

©Bristol Press Ltd. / Noah's Ark

the jaws of the family cat. The cat's owner found the little rat terrified and half-dead in the cat's basket and rushed her off to a nearby vet. Her wounds were largely superficial and having given the vet a large bite, she slowly began to recover. She soon found herself placed in a carrying box and was off on her way to The Rat Rescue Centre at Noah's Ark.

Bev immediately took pity on their latest new arrival. She felt the rat was unsuitable for immediate re-homing, having attacked the vet and since the rat was still in a state of shock, she decided to take her home and try to rehabilitate her.

'This young rat was probably about four-months-old when she came to us,' says Bev. 'She is a brown hooded white rat and seemed friendly and I decided to keep her if the owner did not come forward to claim her. My house is already home to Piper Rat, who was found up a drainpipe and Hans Rat, found in a gutter in the middle of Bristol. Noah's Ark is good at caring for all types of small, unwanted animals, not just the cuddly ones. I christened my new rat "Alice – Alice in Wonderland". I can only wonder about her past.'

OLIVER TWIST
THE DAPPLE GREY PONY

Born with a twisted face and destined for slaughter,
REDWINGS HORSE SANCTUARY offered him
a safe, secure future.

In April '97, a dapple-grey pony found himself at a horse sale in Nottingham. He was quite thin and not at all handsome. Because of his twisted face, he was unable to wear a bit and bridle and therefore could not be ridden. He was at the horse sale to be sold for slaughter.

But luck was on his side. As soon as she saw him, animal lover Beverley Barrett, just knew that she *must* intervene. Because of his twisted face, she named the pony Oliver Twist – soon to be shortened to Ollie.

A local paper in Nottingham, picked up on Ollie's story and cash soon flowed in from the public to pay for his vet's bills. But Beverley knew she must find a permanent home for her Ollie where he could receive expert care.

She contacted Redwings Horse Sanctuary, where she knew they wouldn't judge him just because he didn't look like other ponies. They accepted him without question and after a fond farewell, Ollie was off to his new home in Norfolk.

'Ollie has settled in well,' says a Sanctuary worker. 'We realise now he was probably born with his twisted face disfigurement but he experiences no discomfort from it. Although his face cannot be straightened, he will require extensive dental treatment to prevent his teeth causing damage to the inside of his mouth.'

'Ollie is a perfect example of why sanctuaries like Redwings are needed. He is not in pain or suffering but he is deemed to be of no use because he is unfit to be ridden and therefore unproductive. It is a sad reflection on society that only the physically perfect in the equine world are sought after and given the chance of life. At Redwings, we make no demands of Ollie but allow him to run free with his friends and enjoy the rest of his life in safety.'

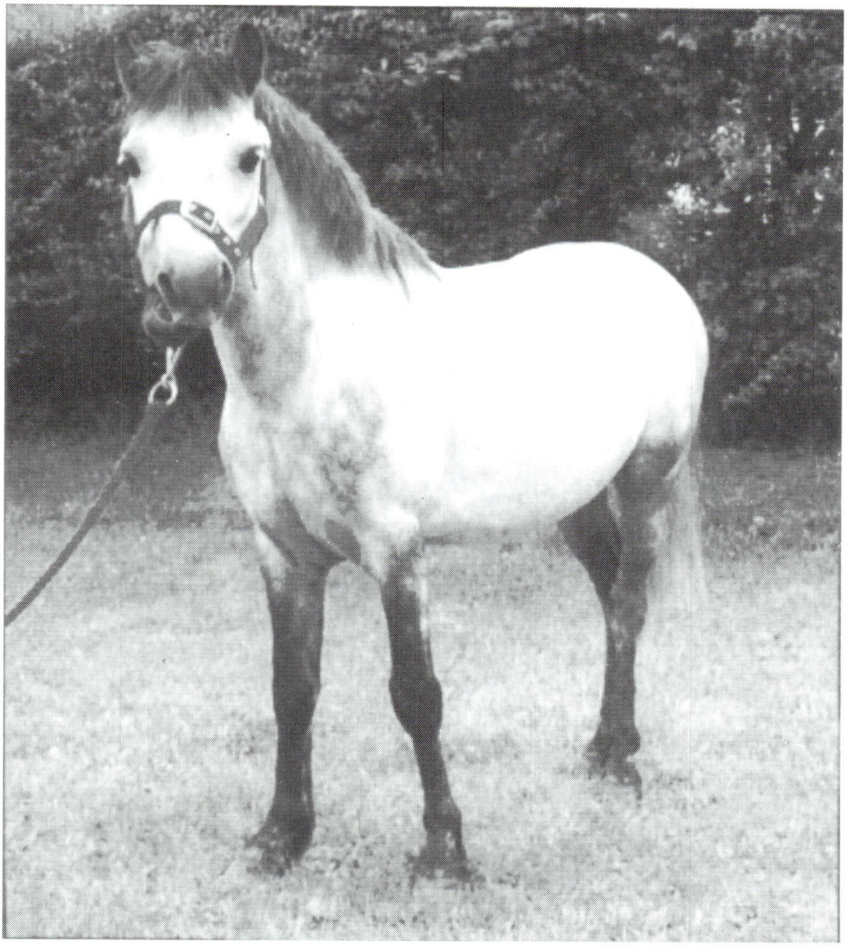

©Redwings Horse Sanctuary

OLLIE

©Redwings Horse Sanctuary

OLLIE : *Content at the Sanctuary.*

Much to Ollie's enjoyment, the fact that he was different meant he received more than his fair share of attention! Not only was he a star in the papers, he also appeared on video and joined Redwings' adoption scheme to become one of the ponies that supporters are able to adopt.

Redwings cares for over 1,000 horses, ponies, donkeys and mules located across four farms. Their aim is to rescue these animals from neglect and slaughter and provide them with a caring home for the rest of their days. They make sure that animals that arrive together as friends, stay together and are never split up. They investigate all the cases that are reported to them, relying heavily on a nationwide network of over a hundred voluntary welfare officers.

‘Some rescues arrive at our Sanctuary through no fault of their own, frightened and so mistrustful of humans that we cannot begin to imagine the terrible experiences they must have suffered,’ says Mark Hudson, one of the Sanctuary Managers. ‘They arrive unwanted, labelled as dangerous and with no place in equine society. With patience, understanding and humane handling, we can rebuild the trust that has been destroyed. Their life becomes a joy instead of a battle.’

Visiting the Sanctuary is a unique experience, not just for horse lovers but for anyone interested in flowers, birds and the British countryside. At Redwings, there is a special Pets Garden of Remembrance, where owners can lay a plaque in memory of a beloved pet. It is a tranquil area, bordered by an abundance of shrubs, mature blossoming trees and a spectacular display of flowers and rose bushes. Special message plaques nestle amongst the flowers in memory of special animal friends.

Redwings is host to a vast array of birds. Jackdaws can often be seen perched on the backs of horses, plucking out loose tufts of hair to line their nests! Chaffinches, Goldfinches, Sparrows and Pied Wagtails are all in abundance gathering pony hair for their nests. Starlings are more often than not found in the fields, foraging for insects where the hooves of horses have stirred up the ground.

Pheasants can be seen sharing food with the horses which has spilt from their feed bins.

As the staff at Redwings point out : ‘Our rescued equines all give an unintentional helping ‘hoof’ to many of our British birds!’

YORKIE

THE GERMAN SHEPHERD

A Bag of Bones who, with the help of PINE RIDGE DOG SANCTUARY, became the best companion anyone could ask for.

When Connie Cuff, Principal of Pine Ridge Dog Sanctuary, collected Yorkie from Wales, he was, in her words : 'A pathetic bag of bones, in a terrible state, with no fur on his tail, which for amusement he continually chased and chewed.'

'He was about ten-months-old, dejected and nervous and had no idea what love or fussing was. He had been shut in a shed and had never been taken for walks. He had never been shown any affection, yet had so much to give.'

Connie felt the best cure for Yorkie was to find him a *very special* home as soon as she could.

About this time, Bill Gray from Slough, was looking for a companion – a dog with whom he could build a close relationship.

Says Connie : 'Bill's heart instantly went out to Yorkie. It has turned out to be the best thing that could have happened for Bill. Having Yorkie has made a world of difference to him and together they have made so many more friends.'

'Yorkie is a different boy now and has truly blossomed into a beautiful dog. He has a very special relationship now with his doting master. He loves his 'Dad', loves his home and wants for nothing.'

'An example of how, given the chance, both love and companionship can be brought to both dog and owner.'

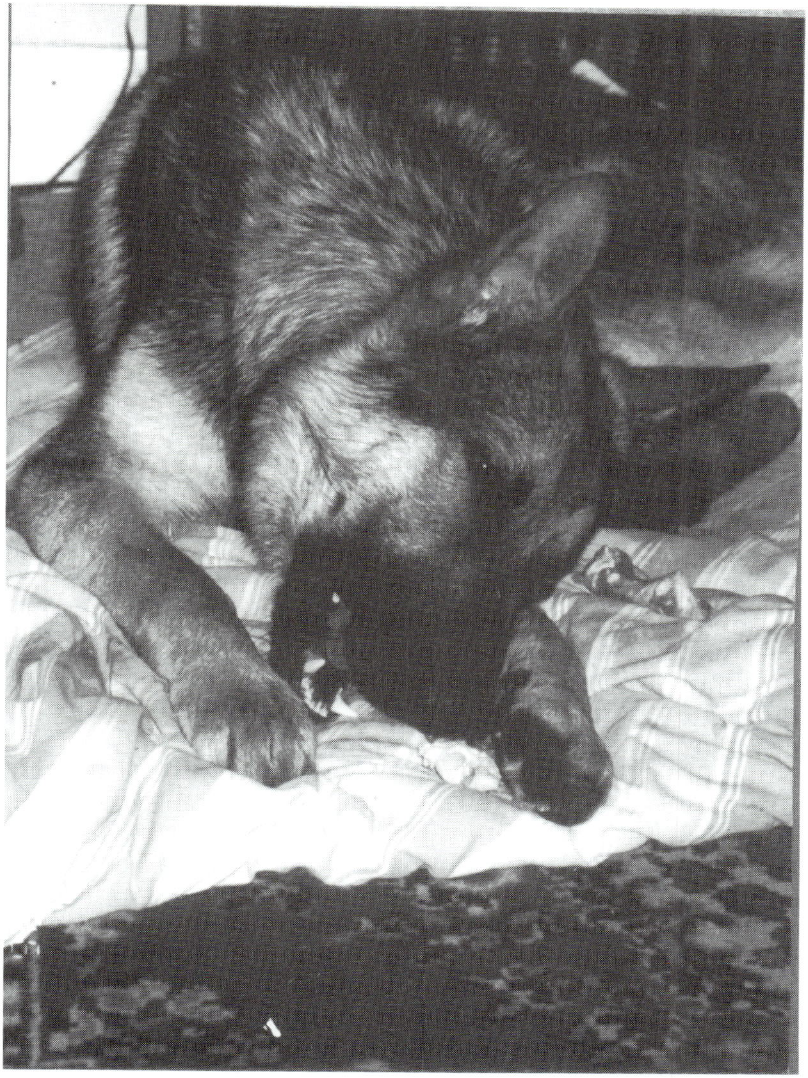

YORKIE : *Loves his Dad, loves his home and wants for nothing!*

ELAINE : *With her Hedgehog patients.*
An Albino Hedgehog hides under her elbow.

NELSON
THE HEDGEHOG

HEDGEHOG CARE came to the rescue and got more than they bargained for!

A visit to Hedgehog Care at Authorpe – a humble, unconventional Hedgehog Hospital – will make you raise your eyebrows and smile.

Within the little cluttered cottage, you will find the Olive Brown Ward (named after a long-stay brain damaged hedgehog), items of hedgehogabilia for sale and upstairs, to aid funds, a 'Hogsfam' clothes sale! You will be in a world full of hedgehogs. Elaine Drewery and Nigel Brocklesby hope you will leave so much more aware of the many dangers facing these loveable creatures.

'So often the casualties are orphaned, injured, poisoned, ill, cold or starving and need help,' says Elaine. 'If a hedgehog doesn't bother to curl into a ball, he is not tame but ill. If he is lying on his side, his condition is serious.'

'Although it is our ultimate aim to release the hedgehogs back into the wild, not all patients make a full recovery so a few become permanent residents and need a sanctuary for life to cater for their disablement.'

Elaine's life with hedgehogs has been full of humour. 'It's been a joy! My humble facilities are extremely "kitchen sink". I pay for medicines and treatments myself which explains why my house is falling down for want of repair and I walk around in rags! I can accommodate over one hundred patients at a time (generally there are about twelve resident hedgehogs to look after), I'm open twenty-four hours for treatment and all this is mixed in with running a Post Office, dog clippings and portrait painting!'

'On my son's wedding day I wore a posh hat as I syringed liquidized cat food into Impey hedgehog's little mouth for her hourly feeds. At the same time, boxes and buckets of hedgehog casualties from all over the area mixed with the confusion of the wedding guests!!'

Elaine remembers each and every rescue. 'First there was Wilfred Pickles, a baby hedgehog I found in a stable. He stayed over the winter and he taught me a lot before his release in May. Anything that remotely smelled of burning, such as coffee, ash and Jeyes fluid sent him into raptures. He would lick the object, foam at the mouth and flip the bubbles onto his prickles!'

'Then came Spike, also an abandoned baby. Having a poor spine, a local lady gave him laser treatment and taught me how to exercise his back legs. He would lie contentedly whilst I worked his limbs to the tune of 'The Day We Went To Bangor'! He was not destined to return to the wild but he taught me so much about hedgehog communication.'

'Miss Vincent arrived with burns and gangrene in all her feet; Hatti from Trusthorpe was too small to survive hibernation so she stayed for the winter; Colonel, happy to stay, is very welcome.'

'Lady Rambo has a serious skin complaint and despite several attempts to release her, she seems to know she would not survive without regular immersions in my bucket of Alugan solution. She is gentle and perfectly mannered and ideal for school visits, 'Save The Hedgehog' campaigns and P.R. exercises.'

Then there was Nelson – soon to be known as Nellie! This little hedgehog came from Frisby and immediately gained the name Nelson as he was suffering from a very bad eye injury. The eye was too badly damaged to save but Elaine dished out daily puffs of antibiotic powder and with patience and time, a plump little hedgehog emerged, with one bright, beady eye and, on the other side, one clean, tidy socket.

As he had not been captive for too long, adaptation for release back to the wild was not complicated. Elaine left a nightly supply of cat food near his release box until he remembered that finding his supper was once more his own responsibility. Elaine then contacted Nelson's rescuer, who had asked to be notified when he was ready for home. It was arranged for Nelson to return home the following weekend.

At 5am. the next morning, Elaine heard agonizing screams from inside Nelson's nest. Thinking something was very wrong, she snatched him up and ran panicking upstairs to rouse someone to dash her off to the vet for help. Any squeaking from Nelson quickly drowned the groans of resistance from a sleepy human!

'Then, quite suddenly,' remembers Elaine, 'out popped a tiny head. A little hoglet had been delivered!'

Trembling with excitement, Elaine hastily put Nelson (who had now quickly become 'Nellie'!) back into her nest and prayed that her own interference had caused no harm. She kept vigil, listening for the moment when she could distinctly hear the sound of sucking.

'It was three weeks before I was able to count seven hoglets. I was as thrilled as any Grandmother!'

All the little hedgehogs underwent the Hedgehog Care routine of hardening off in outside runs to accustom themselves to the weather. They were nine-weeks-old when, with Nellie, they left for a life back in the wild, having been reared without the perils of predators and with the benefit of a constant food supply.

Many, many hedgehogs continue to fill Elaine's life. 'My grown-up family is gradually moving out – probably because there are hedgehogs here, there and everywhere! The little creatures cast their prickles ready for unwary bare feet on the carpets. It is painful to tread on a prickle but the resulting energetic leaps and howls of pain never fails to appeal to my sense of humour! In the chaotic untidiness of my house, an occasional garment dropped carelessly on the floor goes unnoticed until some delighted hedgehog is only too pleased to snuggle into a sleeve for a cozy snooze!'

'Please remember,' says Elaine, 'please have a heart and pull a bonfire to pieces before setting it alight. When driving, please avoid hedgehogs when possible and pick them up if they aren't squashed. They are quite likely to survive with prompt attention. It is a lovely feeling to pull a hedgehog back to life. During their absolute dependence on me, they are very affectionate. I have no veterinary qualifications but if people bring them, then I will do my best with love.'

'To know hedgehogs is to become besotted!'

PRINCESS LEIA
THE CHEETAH

How CHEETALERT!, together with THE CHEETAH CONSERVATION FUND, are fighting hard to save the magnificent Cheetah from extinction.

The Cheetah has had a close association with man for the past 5,000 years but now man is the only animal that can save it from extinction. Cheetalert! has been launched to prevent this beautiful animal becoming just a memory. It is the only project in the United Kingdom and one of the very few worldwide, dedicated to helping this animal win the race for survival.

The Cheetah (or 'Chita' meaning 'The Spotted One'), a unique and specialised member of the cat family, is generally found on open grassland but will also inhabit dense bush and even mountainous terrain. They are known to be the fastest of land animals, able to reach speeds of 40 mph. within three strides, with a top speed of 70 mph.

They are currently under increasing threat. The loss of even a few animals could spell severe consequences for the remaining population.

At the turn of the last century, there were about 100,000 cheetah spread throughout Africa and Asia. Their numbers have now dropped to around 9,000, with some estimates put as low as 5,000. Their genetic problems give rise to sterility, birth defects, low resistance to diseases and an inability to cope with environmental change. In some African countries, with a thriving tourist industry, hundreds of vehicles harass the cheetah and little do the tourists realise, that by their very presence, they could unwittingly be responsible for the gradual decline in cheetah numbers. Cub mortality is extremely high – only one in twenty cubs actually make it to adulthood.

As David Nelhams, Founder of Cheetalert!, points out : 'The cheetah, now an endangered species, needs *all* the help we can offer.'

©Melvin Nelson / Cheetalert!

The Race for the Cheetah's survival goes on.

© CCF / Cheetalert!

Daniel & Laurie Kraus in Namibia working on a cheetah.

Cheetalert! is closely affiliated to The Cheetah Conservation Fund (CCF). The Founders of CCF, Daniel Kraus and Laurie Marker-Kraus, are totally dedicated to a most successful conservation effort. Laurie has worked amongst cheetah since 1973 and is considered one of the world's leading authorities, raising a cheetah from birth and teaching it to hunt on the plains of Africa. Daniel too, has extensive experience with many species of Big Cats.

In 1991, they sold the property they owned in Oregan and raised enough money to move to Namibia – home to the largest population of free-ranging cheetah – where they now devote their lives to the fight to save the cheetah. There is no receptionist, no carpeting, the electricity is diesel powered and the telephone is a party-line shared amongst eight neighbouring farms. But this is The Cheetah Conservation Fund headquarters, situated in a remote town in a sparsely populated corner of Namibia. It is probably closer to the animals it protects, than any other international wildlife organisation.

The Namibian Government has given Laurie and Daniel full clearance for their research and recently they have begun proceedings to purchase land in Namibia for a Cheetah Conservancy. Funds for this will partly be raised by the supporters of Cheetalert!

Most important of all, Laurie and Daniel are winning over the support of local farmers who, when they trap a cheetah, will now consider calling CCF instead of automatically shooting it.

One such cheetah, given her chance, was Princess Leia.

During the January rains, a female cheetah was moving through one of Namibia's large farms, accompanied by her two ten-month-old cubs. A farm worker, near one of the cattle posts, spotted them and attempted to kill the mother. She saw him approach and started to run away, but not before the worker managed to get close enough to spear her. Due to the rain, he was unable to track her, but it was assumed she died because she was severely wounded.

Shortly after this incident, a neighbouring farmer started experiencing problems with two very young cheetah catching his goats. These cheetah were very inexperienced and chased the goats around for some time. The farmer shot one of the cheetah but decided to trap the other one because it did not leave the area as expected after it's sibling had been shot.

This young cheetah was near to starvation and in very poor condition so the farmer called the CCF.

Daniel and Laurie worked hard to bring her around, back to good health and in doing so, to some extent she became socialized. Everyone at CCF became very fond of her and she was named Princess Leia. She now happily resides in their large compound next to Chewbaaka, their resident Cheetah Ambassador.

David Nelhams points out why Princess Leia's story is so *very* important. 'It is probable that the two young cheetah were the cubs of the mother that had been speared. Being too young to hunt wild game properly, they resorted to catching 'easy prey' such as goats. As a result of their mother being speared, two 'problem' animals were created. Every year, cheetah are trapped and killed simply for being seen on a farmer's property, whether or not a problem actually exists. By randomly catching any predators, family units are broken up and social behaviour disrupted. This particular incident resulted in the removal of three female cheetah from the wild. Their population is already genetically compromised. They can ill afford to lose such valuable genetic stock.'

'For the cheetah, the race for survival goes on. CCF and Cheetalert! are doing all we can to help this magnificent animal outrun man's 'progress' before it is too late.'

© Cheetalert!

PRINCESS LEIA : *Her family unit broken up.*

FELICITY

THE HEN

With the help of CHICKENS' LIB, she exchanged the gloom and torture of the Battery House, for a life pecking free in an orchard.

FAWN – the Farm Animal Welfare Network – formerly known as Chickens' Lib, campaigns tirelessly to outlaw the barbaric battery system for laying hens. They continually witness the horrendous practices in factory farming and long for the return of the sight of more free-range hens pecking outside in the cool, fresh air.

Clare Druce, who, with her mother Violet Spalding, founded Chickens' Lib, says : 'There are around thirty-million battery hens in the UK. The terrible sights we come across form merely the tip of the iceberg. We see sheds often containing up to sixty thousand hens. Whereas on free-range it is easy to detect a weak hen, in the gloom of the battery houses, the bottom cages are almost in darkness. Who sees? Who cares? The battery hens suffer in their cages until they collapse and die.'

'We have been into many battery houses, places of sickening cruelty and horror. Many crimes are being committed on the factory farms and also at the slaughter houses. We have bought many end-of-lay birds. They have had tumours, been crippled, even blind. The feet of battery birds are frequently damaged. Most are in a featherless and pitiable state, their breasts rubbed raw and bare of feathers from the cage bars. In their frustration, they have pecked at each other – there is nothing else to do in the long days in their cages.'

For over a year the hens have known nothing but the confines of their metal cage. A typical cage measures 18" x 20" housing five hens (each with a wing span of 30"). For months on end, they will have stood or crouched down on the sloping wire floor, laying their eggs, their only movement being the stressful competition to reach food and water. No litter, no earth, not a wisp of straw.

© Bob Allen

Hens Behind Bars.
'When will the nightmare end?' asks FAWN.

© Chickens' Lib

Happy, Healthy Hens.
Originally these hens were found in a featherless condition, in the cruel confinement of battery cages, by Chickens' Lib.

Some pace anxiously for an hour or more, often climbing over each other, in a hopeless attempt to find seclusion for egg laying. After about one year in the laying cage, the hens go for slaughter.

'Yet within twenty-four hours,' says Clare, 'rescued hens can construct perfectly formed nests, pick up the pieces of straw in their often disfigured beaks and carefully arrange the strands to build a protective shelter for their eggs. Recently, a hen snatched from a slaughterhouse, almost disappeared from view in her deep, deep nest. Perhaps she was making up for a lifetime of deprivation?'

'Ex-battery hens are often weak – we have watched a newly released hen teach herself to walk – but after a few days of care, they invariably, and for the first time, exercise their age-old inherited instincts. In their natural state, hens are active from dawn to dusk, scratching around in the straw for titbits, spreading their wings, running, dust-bathing. These are all behavioural patterns denied to the battery hen.'

Felicity, was one of the very few lucky ones when, one autumn, a small group of supporters from Chickens' Lib, travelled to a village in Derbyshire, hoping to buy some end-of-lay hens.

The first farm they approached had nothing to sell but the owner directed them to a neighbouring farm, having twenty thousand hens, owned by three brothers.

'We looked for the owners but no-one was about,' says Clare. 'The door of one of the sheds was open so we had a look inside.'

'The hens were in an appalling state, many of them almost featherless. The stench was unbelievable, so bad that even we, seasoned battery visitors, could not go inside. At the doorway was a mask, obviously a vital part of equipment for anyone who had to enter the unit and breathe the foul air. Also in the doorway were the decomposed remains of several hens.'

'We looked into another ramshackle and dimly-lit building, where thousands more hens were imprisoned.'

Totally appalled and disgusted, the group were unable to get help from the local police or the Ministry of Agriculture, so they returned in the afternoon to have a word with the owners. They were still determined to save *some* of the birds.

Without warning, an Alsation, which had been chained up earlier in the day, bounded out of the yard behind the units and bit Violet on the legs. She still bears the scars. 'We beat a quick retreat and spent the rest of the day in Casualty!' says Clare.

About a fortnight later, a further group of supporters attempted to buy some of the hens and managed to come away with twenty-four, paying £12 for them.

'The hens were delivered to us the same day,' says Clare. 'Many were featherless. They looked grotesque with their lack of plumage, white combs and overgrown claws. One had growths on the side of her face and some were very thin. They were unused to any real human contact and were in a total state of terror.'

Chickens' Lib did their utmost to bring about a prosecution against the battery owners. They treated the new arrivals with an abundance of love and within a few weeks the pathetic victims started looking like real hens. Most went to good, new homes to experience the feel of fresh air and sunshine.

Felicity, together with three of the other hens, stayed in the care of Chickens' Lib. She enjoyed the freedom of an orchard, to wander and peck on grass amongst the trees in the company of a dozen or so other hens.

'She was always friendly, extremely tame and endeared herself to us,' says Clare. 'She appeared in a national magazine and graced FAWN's own poster, proving herself to be very photogenic! We had given her the name 'Felicity' as this means 'happiness'.'

When Felicity died, it was far from the horrors of factory farming and the slaughterhouses. She did not have to be 'put down' by a vet. She died peacefully and was buried under a flowering cherry.

'Very sadly,' says Clare, 'the battle against factory farming still is not won. The horrific system goes on. It is consumers that keep the nightmare system up and running. No doubt, a lot of progress has been made. No campaign of this magnitude succeeds overnight, but if we boycott battery eggs and buy genuine free-range eggs, cages will become a thing of the past.'

FELICITY : *On the day she was rescued from her life as a battery hen.*

© FAWN / Chickens' Lib

FELICITY : *The name that means "Happiness". Freedom is running on soft grass with a new, healthy coat of feathers.*

© FAWN / Chickens' Lib

OSCAR & KYLE
THE OTTERS

INTERNATIONAL OTTER SURVIVAL FUND came to the rescue for Kyle, found with a plastic ring tightening around his neck and Oscar, the otter cub lost and alone.

International Otter Survival Fund is the only organization in the world dedicated to the conservation of all thirteen species of otter. Their work involves actively protecting these endangered species, safeguarding areas of good habitat and supporting people working in research and rehabilitation worldwide.

'Otters are threatened by pollution, loss of habitat, human encroachment, farming, roads and fisheries,' says IOSF. 'Many other species have vulnerable populations which could eventually become extinct if we don't act to protect them. The Marine Otter, the Giant Otter, the Southern River Otter are all classified as 'endangered', while the Hairy Nosed Otter and the Spotted Necked Otter do not have sufficient data even to classify their conservation status. The more that is known, the more it may be possible to help otters to survive and perhaps reintroduce them into habitats they have now deserted.'

'Protect the otter and you protect fish, frogs, invertebrates as well as water meadows, rivers, estuaries, lakes and coastal zones. A good otter population indicates a clean, healthy river system and that is what we all want.'

On the Isle of Skye, IOSF have developed a rescue centre and habitat where they continually care for injured and orphaned otters. Many road casualties are reported to the centre each year, which has resulted in testing reflectors in some areas to help keep otters off the roads.

Other otters come from further afield and trying to get casualties to the rehabilitation unit on Skye can often be a problem, particularly in winter, because of ferry connections.

But all new arrivals at the hospital are given a thorough examination and are cared for as long as is necessary in the croft pens. Many are then re-released back into the wild. As far as possible, cubs are put back to the same area where they came from. Many are monitored after release and return for a while for food but very soon they will become totally independent.

Oscar and Kyle are a couple of otters who were to find themselves in the caring hands of IOSF.

Oscar was the little otter found in a field in Shropshire, wandering all alone. He was taken along to the Vale Wildlife Rescue Centre but since they do not have long term facilities needed to rear a little cub, and since it was far better for him to be reared with other otters rather than on his own, an appeal went out on Radio 5 Live! Oscar was soon to embark on his travels.

Aboard his own aeroplane, Oscar headed North to the IOSF hospital on Skye. However, the flight did not go totally as planned. Grace Yoxon, Director of IOSF, continues : 'Oscar took off from Staverton Airport in Gloucestershire but driving rain and winds forced the pilot to turn back from Skye. He was diverted to Inverness Airport, one hundred miles away! We set off for the two-hour journey straight away and returned later that night with the little traveller.'

Safe on the Isle of Skye, Oscar is proving to be a very comical, energetic little character. After a twelve-month stay, he will be ready to return to his life in the wild back in England.

© IOSF

OSCAR : *Feeding time for the lively little cub.*

One of the many problems for the otters on Skye is litter. Kyle, a young otter discovered in Kyle of Lochalsh, was found with a plastic cable tie around his neck. As he grew, so the plastic became increasingly tight.

'The local fishermen had been marvellous in trying to catch Kyle, which is not the easiest thing to do,' says Grace. 'Eventually they managed to lock him in the wheelhouse of a boat and we were called. The semi-dark, confined space of the bottom of a fishing boat is not the best place to catch a reluctant otter but after some time we had him safely in a cage.'

'The plastic strap had cut deep into his neck, his breathing was poor and his head was swollen quite alarmingly.'

Kyle soon made a good recovery and was successfully released back to the harbour at Kyle of Lochalsh. He was well known here and this was obviously *his* territory. Once the door of his release cage was opened, he swam straight towards his usual hiding place under the pier. The local fishermen kept an eye on him and two weeks after his release he was spotted foraging about three miles away from Kyle, going back and forth to his regular haunts.

'Kyle's story,' says Grace, 'shows us yet another of the dangers caused by man. Litter – plastic ties can be fatal to the otter. If we had not got to Kyle when we did, the fishermen were going to have to shoot him to put him out of his misery. He was in such a bad way.'

Thanks to IOSF, Kyle once again continues to splash and swim freely in his own waters.

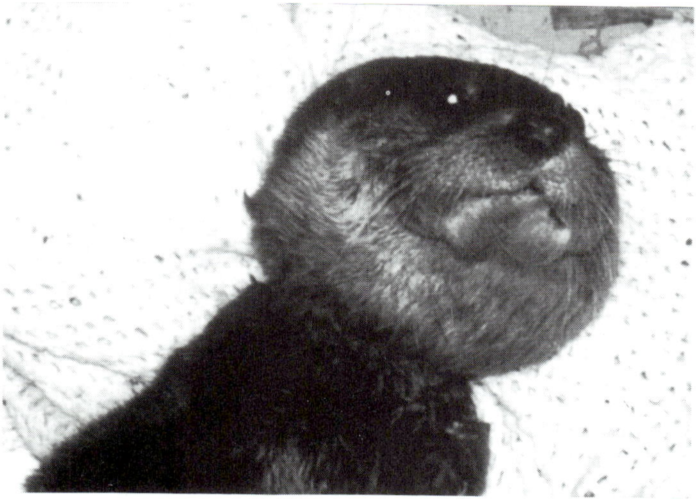

© IOSF

KYLE : *With plastic tie tightening around his neck.*

© IOSF

KYLE : *Recovering and almost ready to return to his favourite haunts.*

BEAUTY
THE PIGEON

Not 'Only a pigeon' but a member of the team
at PALS ANIMAL SANCTUARY.

'About fifty rescued rabbits of all colours doze contentedly in the sun. They have previously only ever known life in a hutch. Here they hop unconcerned from one patch of grass to the next. The high fence keeps them safe and they run free in their own self-made warren. They have such a lovely time – it's wonderful to see them skipping, hopping, chasing and playing with other equally contented bunnies.'

Lorraine Drake is describing the rabbit warren at the Protection of Animal Life Society – PALS Animal Sanctuary in Wiltshire. Every variety of creature can, at some time, be found here. At PALS there is no discrimination between animals and 'so-called pests'. They take in *all* living creatures which is how a certain baby pigeon found a very special friend.

One spring morning, a little bird fell out of her nest. This ungainly, pink bundle of flesh, with not a feather in sight, had only days before pecked her way out of a small eggshell into the outside world. Fortunately, a kindly gentleman was passing at that precise moment and soon had the little bundle cradled in the palm of his hand. He looked around but unable to spot the nest, quickly took her to his car. The car was full of noisy, barking dogs being transported from dog pounds in Wales to PALS Animal Sanctuary but this time there was to be one extra little passenger on the long, tiring journey.

Some hours later, the car drew up at PALS and Lorraine gave a warm welcome to all the excited dogs.

'Oh, by the way, do you think you could do anything for this poor little thing?' said the gentleman just before leaving. He handed the gawky-looking bundle of pink flesh to Lorraine. As if sensing a great event in her life, the little bird chose that moment to open her eyes and look up at her new friend.

Lorraine gave her a little kiss of welcome, wrapped her in some warm, fluffy cotton wool and settled her in a cat basket. After a few drops of watery food, the little mite collapsed into her woolly bed. She had had quite a day!

The days, then the weeks passed. Feathers started to appear. 'White feathers!' exclaimed Lorraine. 'All the pigeons I had reared had had grey feathers. Beautiful, snow-white feathers, so I named her Beauty. It was as if this little soul had been sent to reassure me that there is purity and light in the world. I knew that this little bird, with a huge personality, was very special.'

With daily cuddles from Lorraine, the tiny fragile body gradually took shape. One day with a flutter of wings, she took a small leap. Soon she was fluttering more strongly and was able to make short flights. Before long Beauty was soaring with her companions in the freedom of the skies. Many doves and pigeons had found a temporary home at PALS and subsequently flown away. Beauty joined other birds for a short flight, she ate with them, but then she would fly back to be with her 'mum'.

'I believed Beauty would soon join the other birds for good,' says Lorraine, 'but weeks went by and still Beauty stayed with me! Standing on my shoulder, my hand, my head, in fact anywhere she could find beside me. If I called her, she came. Fearless and brave, I worried greatly for her.'

'Visitors come to the office and Beauty immediately starts to play tricks on them! She'll land on them, dive-bomb them, cooing and making people laugh. A special stand was made for her with it's own water and feeding tray, which she soon adopted. When she has inspected all the papers in the office, sorting them all out, she will then fly back to her perch for a drink or a feed. There is always a red patch on her snow-white feathers – where I've kissed her!'

BEAUTY

'One day there was a nasty accident. I always check there is nothing near the car when I move it but on this occasion Beauty must have been hiding just behind a wheel. Checking in the mirror, I saw white feathers flying in the drive. Absolute horror hit me; I dreaded getting out of the car. I saw Beauty's whole tail on the ground amongst the feathers but no sign of Beauty. In tears, I called out to her and spotted her on the garage roof. She followed me into the office with her lovely tail missing.'

'Beauty survived that day and the next. She was still able to fly a short distance looking *so* pleased with herself! She is just so amazing; she has both guts and determination.'

Beauty's feathers grew back and she soon had a full tail again. She joins other birds for a royal fly-past in the sky but still the office perch is her favourite place. Being near Lorraine is her happiest time. She will join in everything. Sitting on the desk, she pecks at bright pins and paper clips, re-organizing them to her heart's content. Everyone loves her and an elderly couple will visit the sanctuary regularly just to bring her titbits. When a puppy ran up to her for a game, she pecked it on the nose and it ran away looking most surprised! Everything fascinates her; she loves life!

On jumble days, Beauty helps to pack all the jumble on the trailer and has to inspect everything that goes aboard. She flies in and out really believing she is helping. She then sits on the car roof having a ride to the main gate before flying off.

She also enjoys the weekly food delivery. As Lorraine climbs on the van to pass the sacks of feed down, Beauty flies in to help. On one occasion, when all the feed was stacked in the shed, the driver was about to button down the cover when they realised Beauty was still in there, busily pecking and inspecting all the goodies left in the lorry!

'Beauty gives us many laughs. She adores us. We adore her,' says Lorraine. 'Many people will say : "Only a Pigeon – Pigeons are pests!" Beauty makes people change their attitude and stop and think. True beauty is in the eye of the beholder and what a person sees as beautiful, says much about that person.'

LEELA
THE KITTEN

Two dedicated animal lovers travel half way around the world, on their own savings, to rescue one tiny stray kitten from the back streets of India.

In 1995, after months of hard work running PALS animal sanctuary, Lorraine Drake and her friend Janet Oddy, a PALS volunteer, took off for India on a much needed holiday.

They were up and away early on their first trip out, the taxi speeding bumpily along the dirt track roads, heading for Old Goa town. Despite cows meandering across the track, the driver didn't seem to worry at all about slowing down. Suddenly, Lorraine found herself screaming out : 'Stop! Stop! Stop! What's that?' The taxi driver didn't understand. 'Stop!' screamed Lorraine again and the driver screeched to a halt wondering what on earth all the fuss was about!

'We tried to get out of the car,' continues Lorraine, 'but the door was jammed. I kept pointing and shouting madly that there was a kitten in the road. The taxi driver looked at me, most surprised. "You want?" he asked. "Oh, yes!" we screamed. He jumped out, ran back and scooped up the scraggy bundle of fur. "You want *this*, Madam?" he enquired, thinking I was quite mad!'

Miraculously, the kitten had survived beneath all the trucks. She was thin and weak but apart from that was still in one piece.

Lorraine and Janet continued on their journey with the taxi driver on the look out for somewhere to get some food and milk – not an easy task in downtown Goa.

'We stopped at a little shack,' continues Janet. 'Never before had anyone requested food and drink for a kitten but there was great excitement. The local Goans provided us with some milk served in half a coconut shell and Leela, as we had now named her, lapped it up. We continued our journey to the Cathedral of St. Francis Xavier in Old Goa town with Leela nestled comfortably in my camera bag, squeaking away, drawing a lot of attention and interest and having a very nice day!'

© J. Oddy

Lorraine offers Leela a little milk,
served in half a coconut shell.

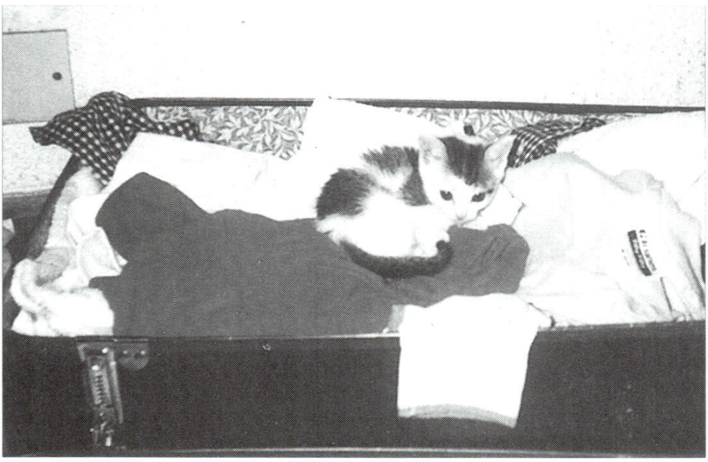

© J. Oddy

LEELA : *Snuggles up in her*
new bed.

When they returned to their hotel, Leela had to be smuggled in – the rules stated quite clearly : No Animals Allowed! Janet raided the restaurant for food and the little kitten, despite being only four-weeks-old, was soon tucking into slices of meat without any problem whatsoever.

Over the next couple of days, Lorraine and Janet made numerous enquiries regarding a home for Leela. They desperately tried to find an animal sanctuary but none of their leads materialised. Then the awful realisation hit them – there was nowhere for Leela to go. The Goans just do not have cats as pets. Life is hard out there for villagers and cats live wild, managing as best they can off scraps. They knew Leela was far too young to survive; if she did she would not be neutered and the whole cycle would go on with more unwanted kittens brought into the world.

'There was nothing else to do,' says Lorraine. 'Leela *had* to come home with us.'

'By this time, our "secret, not allowed in the hotel" kitten was known throughout the complex and by the regular taxi drivers, who delighted in taking the three of us out. But now we needed help. Leela was too young to have a rabies vaccination so she would not be allowed on the flight home and our holiday time was running out. So we arranged to see the hotel manager. He turned out to be absolutely marvellous. He introduced us to his mother, who lived a few miles away and she very kindly agreed to care for Leela until she was old enough for us to bring her into Britain.'

Once Lorraine and Janet were back in Britain, their top priority was Leela. Arrangements had to be made with the Ministry of Agriculture & Fisheries, quarantine catteries and the airline. Numerous telephone calls were made, faxes were sent back and forth to India but nobody could help with regards to the documentation needed for Leela's exit from India. No-one had ever transported a cat out before!

It was to cost a further £3,000 to retrieve the kitten and a huge amount of money for quarantine. But *nothing* was going to stop these animal lovers.

'When you care about animals,' says Lorraine, 'you can't leave them simply because you're on holiday. I cannot stand neglect of animals so we used our own savings to rescue her.'

Three months later, Lorraine and Janet set off back to India. Once re-united with Leela, a never-ending list of arrangements had to be made.

'We were forever negotiating one obstacle when yet another would arise. Firstly the veterinary certificate of health, issued in Goa and needed for Leela to leave India, was not valid in a different state. We therefore had to arrange to leave Goa for Bombay immediately to find a vet there.'

'We booked a flight to Bombay and settled Leela into her flight carrier. To our dismay, we had to hand her over at the luggage gate. As we made our way to the aeroplane and walked across the runway, we kept looking back....... And there she was – on top of all the suitcases bravely looking out and meowing as the luggage tractor trundled towards the 'plane! We saw her safely on board and breathed a sigh of relief. Our homeward bound journey had begun.'

Once in Bombay, Lorraine, Janet and Leela headed straight to Oceanair freight offices and spent two hectic days getting paperwork completed. They leapt from one hitch to another culminating in a mad dash across Bombay to obtain photographs of Leela for her very own passport! They were already planning that maybe they would have to stay behind for a further week.

'But the freight company proved marvellous. All three of us touched down together at Heathrow airport some eight hours later. Once we had landed, we were able to catch a quick glimpse of Leela through a glass screen at arrival quarantine but then it was on to her quarantine cattery for six months.'

'We travelled half way around the world for Leela but it was more than worth it. Three years on, she is now a fully grown, magnificent cat, well settled into her home in the New Forest, living happily with a couple of other cats.'

'Leela is our very *special* blessing from India.'

PAOLA

THE SEA TURTLE

Found trapped in nets and destined for the soup pot, she teamed up with MEDASSET and became a 'star' in their ongoing campaign to save the Mediterranean Sea Turtles.

Lily Venizelos spent her childhood escaping into nature, amongst the creatures on the island of Hydra. She went on to lead a privileged life, marrying into Society and travelling widely. But still she continued to spend hours on the beaches, snorkeling in clear waters and collecting shells from the shore.

By chance one day, whilst aboard her yacht, a rough storm in the Ionian Sea forced her to take shelter in Laganas Bay, Zakynthos Island.

'Here,' remembers Lily, 'I found a Paradise of birds, wild flowers, trees and an enormous pristine beach. I thought I was in a dream. Somehow, I felt there was a message here, one I had to discover.'

Lily returned to Laganas Bay many times. But she was soon to discover 'her paradise' taken over by bulldozers and rows of beachfront hotels. She subsequently found out that Laganas Bay, was a crucial nesting place for the Loggerhead Turtle. This was one of the oldest species of sea creatures and it's very survival was in doubt. Lily knew then that she had discovered the island's message – *Someone* had to save the Sea Turtles. Lily knew it was up to her.

She soon became a determined environmentalist with a passion for sea turtles. She explains : 'The huge sea turtles are beautiful, complex creatures, mysterious enough to become addictive to the biologist and totally absorbing for anyone to watch. They live their entire life, possibly over one hundred years, in salt water but lay their eggs on land. They are the most efficient long-distant swimmers and deep divers known to man. They have existed in the world's oceans for over a hundred million years. Now they are highly endangered and threatened with extinction. We can save the sea turtles, if only we care enough.'

Lily became known as "The Turtle Woman" and before too long had set up MEDASSET – The Mediterranean Association to Save the Sea Turtles.

© L.Venizelos

Lily soon found she was challenging not only the sea-front developers but also sea pollution and the fisherman with his drift net. Her work never stops. Money-raising for her charity is never-ending. She attends International Conferences, holds lectures for school children and students and lobbies Governments. MEDASSET, the only organisation working exclusively on the conservation of sea turtles throughout the Mediterranean, has grown and expanded into research. They study the turtles' nesting grounds and identify the areas where they feed and spend winter months so that they can be protected from fishing activities. They have surveyed thousands of kilometres of coastline, providing invaluable information about the life history of sea turtles. Many are found dead every year due to ingestion of small pieces of garbage left behind on beaches or dumped into polluted seas. MEDASSET have initiated an International campaign to increase public awareness on this issue.

Lily still fights hard to save Laganas Bay. This Bay is one of only a few specific beaches on which turtles will lay their eggs in the Mediterranean. Here, hundreds of turtles emerge annually from the sea, in the silent darkness of night. They drag their huge bodies onto the beach, leaving tank-like tracks behind them, they dig a hole with their back flippers and they lay their eggs. They need undisturbed, sandy beaches to nest and the construction of the beachfront hotels, with their noise and floodlights in the evening, has created a variety of problems for successful nesting. Many turtles too were being killed due to speedboat collisions and, mainly due to MEDASSET, legislation has at last been passed, banning speedboat traffic in the whole of Laganas Bay.

Laganas Bay, once a Paradise Island,
now invaded by tourism.
This is the nesting beach but where can the turtles nest?

The paradise that should belong to the sea turtle is fast disappearing. Lily knows her turtles are an endangered species and their plight is continuing to worsen. She continually welcomes information on turtle sightings, strandings and nestings.

In April '97, a certain Sea Turtle named Paola, became one of MEDASSET's most interesting projects.

An injured sea turtle was found caught up in the nets of an Italian fishing boat. As she was brought to shore, her life hung by a thread, with the choice of a plain, brutal death or straight into the cooking pot for soup.

That night, fate was to smile down on this turtle.

The young daughter of an Italian restaurant owner had just been watching a television programme about turtle conservation and the work carried out by The Naples Institute for Zoological Research. As soon as she saw the injured sea creature, she ran to beg her father to help her rescue it.

Together they drove the turtle all the way to the Research Centre. Under the care of Dr. Flegra Bentivegna, an operation was performed to remove a large number of fish-hooks from her throat and body and the turtle was nursed back to health. In honour of the young girl who had come to her rescue, the turtle was given the name Paola.

Five months later, sea turtle Paola was fit and well and ready to be returned to her natural habitat, the sea.

But one of the projects at the Research Centre, was the monitoring, by satellite, of sea turtle movements around the Mediterranean. Paola was 87cm. long and weighed a massive 50 kilos. It was felt that maybe it *could* be possible for this turtle to carry a transmitter and provide valuable information in the research.

Here MEDASSET stepped in and offered to organise the release and provide the necessary oceanographic information for the area to enable Paola to be tracked. They suggested the island of Kefalonia, where many sea turtles come to lay their eggs in the summer. Here she would be kept well clear of the area off Italy where fishermen still use illegal drift nets and turtles are still illegally taken for food.

A new transmitter was obtained from the USA and in a 12cm. protective case, with a 15cm. aerial, it was attached to Paola's back with glass fibre and silicone.

The day came for Paola to be released. She was to be given a royal send-off in the presence of local and Italian dignitaries, a large crowd of local inhabitants and a television crew who were there to shoot the underwater photos. Paola set off on her travels!

PAOLA : *The Sea Turtle, released into the Mediterranean and tracked via satellite through an electronic transmitter on her back.*

For the next few weeks, Paola meandered her way south-east down the Peloponnese coast. Her exact location could be monitored but what she was up to was a matter of pure speculation!

Then Disaster! Download from the satellite revealed that Paola had ceased transmitting just off Cape Tenaro. What had happened to Paola? Was this expensive project to end so quickly and abruptly? MEDASSET immediately issued press releases, the Ministry of Merchant Marine alerted port authorities and sailing vessels. News bulletins carried the story and updated reports. A full description of Paola was widely circulated – she was becoming a celebrity! Had anyone seen Paola?

Nineteen calls reported sightings of turtles in the Peloponnese area – most of the turtles having been killed by fishermen or hit by boats. But incredibly, the first call came from a man who reported that a fisherman, on a trip in the Aegean Sea, had released a turtle with a transmitter from his nets.

'Yes,' said the fisherman who was finally tracked down. 'A turtle did get caught in my net as I was fishing in shallow water off Koundouros. I took it to the beach and cut my net to free it. It had a transmitter on it's back – I thought it must be important. There was a sheet of plastic entangled around the aerial, which I cleared away. It was fit and healthy and rushed down the beach back into the water!'

Lucky again, Paola had come across a friendly and humane fisherman. Once back in the water, she had resumed transmission from precisely the place the fisherman had released her. A further transmission from Paola came between the island of Lesbos and mainland Turkey.

So, steadily Paola continues her journey, providing much needed information. The monitoring of her movements will benefit all sea turtles with an uncertain future.

On the turtles' behalf, MEDASSET will fight on. 'Even small changes in our attitudes can improve our environment,' says Lily. 'Please help the Sea Turtles of the Mediterranean!'

NOEL / LIAM
THE COLLIE-CROSS

Sad and alone in kennels, desperate for a chance.
NCDL found him the perfect new life.

The National Canine Defence League (NCDL) is the UK's largest dog welfare charity with a network of fourteen rescue centres across the country. More than 10,000 lost, abandoned and mistreated dogs come into their care every year. They truly believe that every abandoned dog deserves a second chance of happiness. That certainly was the case for Noel, a collie-cross.

One Sunday in March 1997, Noel awoke to yet another day at NCDL's rescue centre at Kenilworth. He had been here for over three months and was not happy. Being a working dog, life in a kennel was frustrating and boring. Staff at the kennels desperately needed to find him a working home.

But that Sunday was to be extra special. It was to change Noel's life forever! Roy Dyer, Director of the Essex Dog Display Team, together with Brian Stott, one of the dog handlers, had arrived especially to see this collie-cross.

After spending an hour getting to know each other, the decision was made : Noel should join their prestigious team. And more than that – he could make a guest appearance at Crufts '97 that same afternoon!

Noel left his kennel that day, trotted off to the show ring at Crufts and never looked back. His big day had arrived!

Says Roy : 'It was like being plucked from Sunday League football and being chosen for the World Cup! We're confident that any dog, given understanding and training can be turned around. We changed Noel's name to Liam so that he doesn't associate us with anything that happened to him previously. Oh and we prefer Liam!'

Liam quickly picked up his new skills and is now a fully qualified member of the Essex Dog Display Team. The NCDL found him the best life he could ever have wished for.

© NCDL

LIAM

LIAM : *With Brian Stott from Essex Dog Display Team, and Lisa Tooley, NCDL PR Officer.*

The Essex Dog Display Team go through their paces.

LIAM : *Showing off his new skills.*

Photos © NCDL

ROLLO

THE VIETNAMESE POT BELLIED PIG

The day NATIONAL ANIMAL WELFARE TRUST received an anxious call from a gentleman in London – He had the theatre tickets, the curtain was about to go up but who could look after his newly acquired Pot Bellied Pig?

The National Animal Welfare Trust receives many calls every week to rescue animals in need. On one particular day, a call came through which Franca Cappuccini, NAWT's Welfare Director, will never forget.

'As I picked up the 'phone,' remembers Franca, 'I heard a well-spoken male voice on the line.'

'"*I beg your pardon?*" I found myself replying. "*Did I hear you correctly, or are you joking?*"'

'"*No, No, Not at all!" said the voice. "I can't cope! I don't know what to do! I'm going to the theatre this evening and I can't leave him on his own. I've made a terrible mistake. I should never have bought him. I've rung the breeder. She doesn't want to know!*"'

'This call was not about a dog or a cat,' continues Franca. 'The idiot on the 'phone was talking about a Vietnamese Pot Bellied Pig, which he had purchased for £200 the day before, from an advert placed in Exchange and Mart! It had been delivered to him from an equally idiotic breeder – not only did this guy live in a flat but on the third floor! Can you believe it? The breeder had also told him that this pot bellied pig could be trained to use a litter tray! Some people are very gullible and very, very stupid!'

After Franca had established that the story was true and that, since April 1st had passed, she wasn't having her leg pulled, they discussed the gentleman's predicament.

'He told me he could no longer look after this four-month-old male pig and was *desperate* for help. He had telephoned the breeder he had purchased it from but, of course, she now had his £200 and was not the slightest bit interested in his plight or his pig!'

© NAWT

ROLLO

'Our next step was getting the animal to our Centre as he had no form of transport. He was most concerned that it should be out of the flat by the evening, as he had a prearranged theatre ticket. (I'm not sure if he was thinking of getting a pig sitter, had we not been available!)'

'Once we had the address, off we trotted to a very busy part of North London to collect Rollo.'

'"*Please be very quiet,*" the gentleman said as we entered his flat. "*No-one in the block of flats knows he is here. Please get him out as quickly and silently as possible. I would hate my neighbours to find out. Maybe we could put a blanket over his head?*"'

'At this point, I actually resisted from putting a blanket over the gentleman's head!! I told him that piglets did scream and short of giving Rollo a sedative, which was not possible, a few screams were likely. But he could always tell his neighbours that his dentist had done a home visit!'

'After a few muffled squeals and much laughter, we got Rollo down the stairs and he was off on his way to a new home in Hertfordshire. Here he would be far more content, able to roam freely, not stuck in a flat, getting thoroughly bored, rapidly turning into a couch potato, addicted to 'Neighbours'!'

AFFIE & BETTY

THE FOX CUB & THE FERAL KITTEN

Rescued by the FRIENDS OF THE ANIMALS Wildlife Nursing Unit, these two wild animals soon became two best friends.

In 1845 Queen Victoria purchased Osborne House, a wonderful property enjoying spectacular views on the Isle of Wight. The house is set in many acres of grounds which roll down to the sea shore and it was in these grounds that the Isle of Wight branch of the Friends of the Animals Wildlife Nursing Unit picked up a tiny fox cub.

As Helen Sinclair remembers : 'No more than three-weeks-old, Affie as we called him, had been abandoned by his mother. He was a very sorry sight, ill, covered in ticks and we knew that his chances of survival were low.'

'The following weeks were crucial. We gave Affie around-the-clock care and slowly but surely he grew stronger.'

'Just look at him now! Clearly he prefers to have an easy life and as the average life of a fox in the wild is only seventeen months, can you blame him?'

'This little fox cub has never shown the slightest inclination to seek his freedom. He much prefers to watch T.V. and eat ice-cream!'

'He immediately struck up a friendship with Betty – one of four feral kittens born on a farm. Betty survived due to lots of love and care and occasionally being popped in an Aga to keep her warm!'

'Affie will fall asleep and curl his tail protectively over his best friend. Betty purrs contentedly. Two very happy rescues for Friends of the Animals!'

© FOTA

AFFIE & BETTY :
Two Best Friends.

BUTTERFLIES, BUTTERFLIES, BUTTERFLIES
& SOTIRIS - THE PELICAN

Maggie Avgerinou, of RHODES ANIMAL WELFARE, offers her help to a Pelican and praises the efforts of the people of Tholos, who gave their Butterflies another chance.

The Island of Rhodes – where as legend would have it, a nymph emerged from the deep blue waters and united with Helios – has become famous for it's cosmopolitan air and exceptional beauty.

Tholos, in the North-West, still remains a small, typical Greek village with narrow lanes, low houses and friendly people. Continue for 5km. and you will arrive in the deep, green Valley of the Butterflies.

According to the older inhabitants of the area, the valley used to be named Pelican Valley, after a slave named Pelicanas. (In Greek legend, he was refused permission to marry and therefore committed suicide in the valley). But the name has now changed. The valley has become a butterfly paradise and has been named 'The Valley of the Butterflies'.

An ancient myth states that the butterflies were once the exclusive permanent occupants of this magical gorge. Snakes would wind their way down to the stream to drink the water, then return to the top to hunt. But the snakes soon discovered the butterflies. Their characteristic *hiss* before their attack, became a signal, which the butterflies used as a form of self-defence. With every *hiss* the butterflies would fly away. The snakes therefore had no option but to leave to hunt elsewhere and the butterflies were left to reign in peace over *their* valley.

The slopes of the valley are thickly forested with leafy evergreens, which act as a filter to block the sun. There are carob and wild olive trees, oak, bay and gorse. Thyme and heathers border the valley. Water runs everywhere with the river divided by small rushing waterfalls, creating a valley of low temperatures and high humidity throughout the summer.

During the morning, when the temperatures are lower, the butterflies rest high up on the trees. As the temperature rises during the day, they move down to the rocks that punctuate the stream.

This tiny valley attracts the Panachia butterfly, a butterfly with a strong, red inner wing, which during the summer flies from every part of Rhodes to be present at the 'gathering of the species'.

In July/August, millions and millions of butterflies descend and the whole valley is a vast phantasmagoria. Nowhere else in the world is this species of butterfly found in such large numbers.

But not long ago, the butterflies were NOT happy.

Maggie Avgerinou, of Rhodes Animal Welfare, remembers that it was only the wonderful effort from the concerned villagers of Tholos, that managed to save the butterflies and give them another chance of survival in their magnificent valley.

'This valley,' says Maggie, 'was a special place on Rhodes. At the top of the valley, the reassuring face of the Virgin Mary smiled down from the Monastery of Kalopetra, protecting her enchanted valley. It was where thousands of butterflies would meet during the heat of the summer. They would nestle all together in the cool shade provided by maple and balsam trees and relax in the humidity of the gurgling stream.'

'But things were soon to change. Hoards of tourists, from all over the world, began to descend on the valley to see the butterflies. But these tourists were not happy. These were brown-coloured butterflies with a cream stripe, which meant they would blend into the landscape. The tourists had imagined fluttering, colourful butterflies. They wanted to take good photos – but these butterflies didn't move.'

'So the tourists devised a plan. They found, if they whistled and clapped their hands loudly, the butterflies would be startled into flight and reveal their magnificent bright red under-wings.'

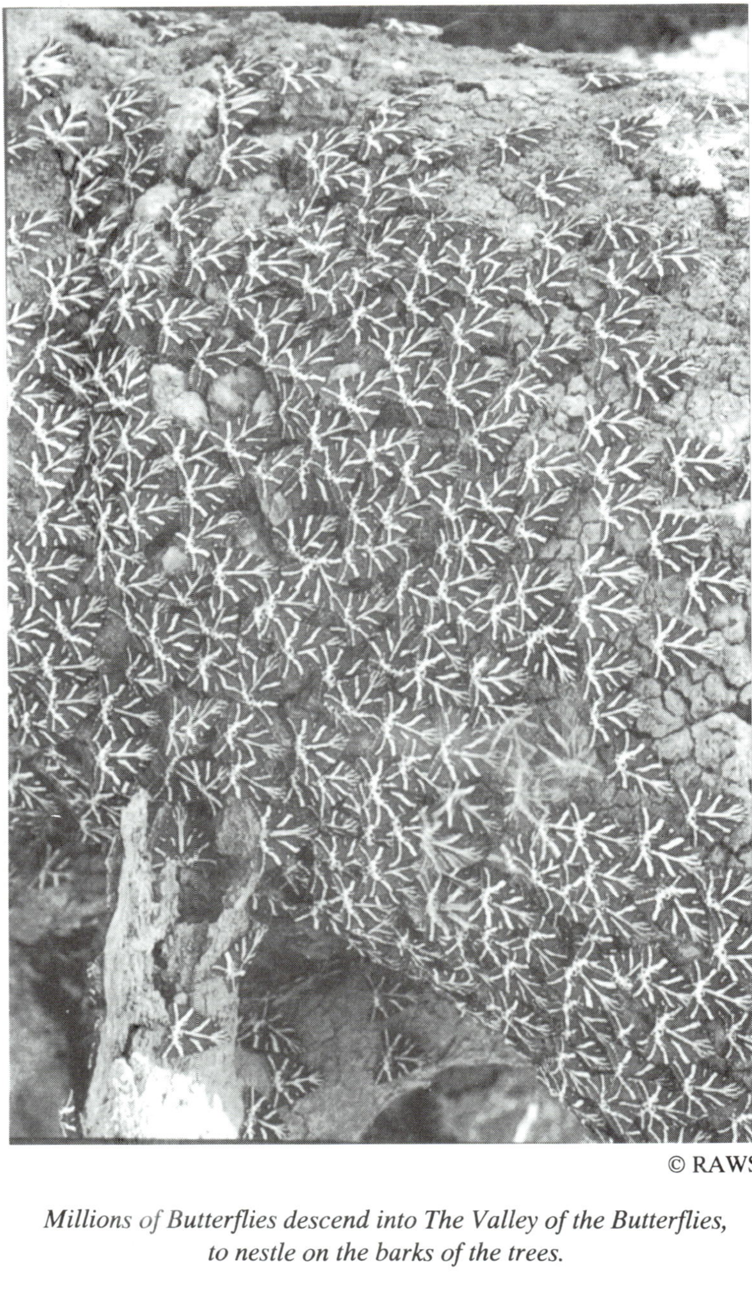

© RAWS

Millions of Butterflies descend into The Valley of the Butterflies, to nestle on the barks of the trees.

'Once flying, a cloud of butterflies would paint the valley red. This is what the tourists had come to see. The tourists were delighted!'

'But the clapping hands created ultra sounds similar to those emitted by bats to locate their prey. The butterflies went frantic. They believed they had become the targets of bats. The inevitable happened. Fewer and fewer butterflies began to arrive in the valley.'

The villagers from nearby Tholos naturally became anxious. They realised that soon there would be no butterflies left at all. It was becoming more and more difficult for their butterflies to remain under the leaf cover. Repeatedly forced into frantic flight, they used up all their precious energy and stored food supply. Many were dying of exhaustion before mating and the others were unable to reach their breeding grounds 30km. away. Before long, there would be a steady decline in the species. It was vital that they remain peaceful and motionless on the bushes and vines, day and night, at the right humidity, thus conserving their energy until the egg laying period.

The villagers held a meeting. They decided they *must* save their butterflies. It was *their* responsibility to protect them.

'Their plan was put in motion,' says Maggie. 'Smoking and noise is now prohibited in the valley. Visitors must keep to the foot-paths. Considerable habitat destruction had occurred due to the 400,000 visitors each year, so new trees and bushes have been replanted. Water from the stream is recycled to increase the flow and raise humidity and water canons have been installed incase of fire. Fences have been erected to prevent goats destroying vegetation and signs have been erected to inform visitors how, by keeping quite, they can help the butterflies.'

Despite a very modest budget, this has been a *huge* success. The butterflies have been given another chance. They show their appreciation by returning in even greater numbers. The European Commission has named this project as one of the most successful examples of eco-tourism. The achievement became a source of pride to the locals and has helped them preserve their source of revenue.

To Maggie, however, the whole project has meant that little bit more :

'Coming originally from Scotland, I have been living here in Rhodes for eighteen years. In that time I have seen a vast improvement in the local people concerning animal welfare issues. Tourists come here and see many strays on the streets and immediately condemn the locals for their cruelty. But it's not that simple. Animal welfare is something very new for them and they are only now learning from the media and from foreign influences. We, at Rhodes Animal Welfare Society, know a vast number of local people, who really *do* care.'

RAWS is a small society, run by Maggie and a few friends. 'All non-Greeks, mainly,' she says, 'because the locals haven't understood the concept of working voluntarily yet. They think we are all mad working for no salary! But we know ideas will change and we are gaining a lot more respect now than in the past.'

Maggie and her helpers daily collect injured, sick and stray young animals from the streets and provide them with medical attention. 'Since Rhodes is a small island, we find it very difficult to re-home dogs, so we work together with some German groups. Many tourists adopt our dogs and we receive many photos of them in their new homes, romping in the snow!'

'We also take in wild birds. Birds are brought to us either having been poisoned or having been shot by hunters. On these occasions, we ship them out to a Bird Sanctuary on the island of Paros, where they are treated and released back into the wild.'

'Pelicans don't usually live here on Rhodes, so when one just appeared in the harbour one day, it caused great excitement! As the Greeks are fond of these birds, all the fishermen were feeding him fresh fish. But they soon realised, all was not well.'

'They contacted RAWS rescue and we immediately took the pelican along to see our local vet. A sight of a pelican was *so* unusual, even a local TV station turned up and the pelican's treatment was shown on TV News several times that day!'

'The vet discovered that there was a very long rip in the skin of his pouch.'

SOTIRIS : *The one who has been saved.*

'This took a while to stitch up as it is very elastic and although he was a very well behaved pelican and very tame, he did have his moments!'

'His stomach was then checked and there was the vet putting his arm right down this very long neck, time after time and bringing out one fish at a time – finally counting up to ten! The tenth and final fish had some fishing line attached to it, all knotted up with at least a dozen sharp fishing hooks. Believe me, this certainly made fascinating viewing!!'

'We baptised the pelican 'Sotiris' meaning 'one who has been saved'.'

'Sotiris went back to the harbour, where he lived very happily, fully recovered but has now moved off to wherever he originally hailed from. I must say, Sotiris is greatly missed.'

VICTORIA
THE GINGER CAT

How an Animal Healer used visualisation to put the spring back in Victoria's paws.

Marilyn Preston Evans is known in her part of Cornwall as 'The lady who helps and loves all animals'. She originally had her own animal and bird healing hospital.

'Visualisation formed a good deal of my healing work, both with humans and with animals,' she says. 'Victoria's story gives others an idea of what they themselves could try to help their animals in some similar difficulties.'

It all started many years ago, when a lady visited Marilyn for healing help for an extremely painful arm. Treatment by Doctors and Specialists had all failed. Later that morning, at 11 o'clock, when she was on her own, Marilyn fell into a form of 'day-dream', mentally placing the arm in a plaster cast. The next day, the patient returned, quite concerned. At exactly 11 o'clock the previous day, she had felt someone 'put a plaster cast on her arm'. In three weeks, the arm was completely healed when the patient had felt 'the plaster being gently removed'.

At that time, Marilyn had never heard of visualisation but has since found it has become a profound part of her healing work. 'Sometimes it is effective; other times it appears not to be so. But with a little stray ginger cat called Victoria, we were lucky.'

Victoria was brought to Marilyn's door with a severely broken hind leg. She had been lying in a garden for three days and the leg was totally shattered. Marilyn's inner voice told her to go to the vet, that she would be told nothing could be done but that she should bring Victoria home and she would be a test-case for animal healing.

Off they went to the vet and Marilyn was told nothing could be done. Victoria would have to be put down. Says Marilyn : 'I replied I would take her home and give her healing and the vet, to his credit, was furious! He felt there was obviously nothing that could heal the shattered leg and it would be absolutely cruel to keep Victoria alive in this condition. He took my name and address and full details about little Victoria and I set off home with the little cat still purring.'

'After ten days of healing and administering Bach Flower remedies – nothing happened. Rather shattered, I wondered what on earth to do.'

Midnight came and, sitting alone in her lounge, Marilyn closed her eyes and began her visualisation technique. In her mind she placed a valuable Chinese vase, shattered in a multitude of pieces. Her thoughts were that she must mend it, with no sign of a break. She visualised a tube of glue, found each piece to fit and gradually put the vase together again, *exactly* as it had been before it was smashed.

Next, she visualised Victoria's shattered bones, some 'bone glue' and very carefully found each jigsaw piece of bone, placing it in position with the glue until the whole of the bone structure of the leg was complete. She then visualised a bandage and carefully wrapped it around the leg and finally 'put together' all the muscles and nerves (although she knew nothing of anatomy!). When she could see the leg perfect again, off she went to bed.

Meanwhile, Victoria was curled up fast asleep in the bathroom – the only room without birds and animals at the time!

In the morning, as the bathroom door was opened, a purring Victoria raced past Marilyn, down the stairs and sat waiting for her breakfast in the kitchen! Her leg was perfectly mended.

Marilyn and Victoria's story, verified by the vet, was published in one of the national newspapers and brought almost three hundred requests for Marilyn's healing help (for which there is no charge).

'The mind is very powerful,' says Marilyn. 'It never loses it's creative force.'

FAZIR, NASSELL & HANEN
THE TIGERS

Along with Six Lions, Two Horses and a Foal,
Seven Dogs and a Python named Murphy! – all rescued
from an Egyptian Circus by ANIMAL DEFENDERS.

In a dramatic rescue mission, the Animal Defenders – an International Animal Welfare and Conservation Group and the foremost campaigners against the use of animals in circus acts – joined forces with the Conservation Group Tusk Force and a local Animal Rights Group in South Africa, to rescue every animal from The Akef Egyptian Circus, stationed in Maputo, Mozambique. All the animals had been abandoned to starve to death in their cages. The Animal Defenders moved in to save three tigers, six lions, a stallion, a mare and her very young foal, seven dogs and an African rock python.

For several years, the Akef Egyptian Circus had gained notoriety throughout Africa, dogged by allegations of welfare neglect. As the circus entered and left various countries, discrepancies in species and numbers of animals began to appear.

In 1993, CITES – the Convention on International Trade in Endangered Species – were informed that Akef's Circus had travelled from Egypt to Kenya with two performing chimpanzees, without presenting any CITES documents. The authorities were urged by the Kenyan SPCA to confiscate the animals on welfare grounds. Nothing was done. The circus moved on.

In Uganda, a further four baby chimps were purchased. This time the Ugandan Wildlife Department managed to confiscate the chimps and the circus was instructed *not* to leave the country. Akef, and his circus, vanished.

The following year, they reappeared in Kenya and further chimps had been purchased with no relevant documentation. Diplomatic protocol impeded any action being taken and the circus set off to Zambia, entering the country once again without complying with Zambian import procedures. The Zambian

Minister of Tourism insisted the circus, which by now had gained a tiger, more chimps and lions all without CITES permits, comply to legislation.

The repeated failure to have the correct permits and paperwork and the fact that the circus was arriving in a country with certain animals and leaving with different ones, led to the allegation that it was simply a front for illegal animal trading. For example, Akef had applied to import seven lions and nine African Grey parrots and had left with six lions and no parrots.

The circus entered Zimbabwe and a Director of Veterinary Services investigated. His report stated : 'The conditions under which the lions and tigers are housed are in gross contravention of the Prevention of Cruelty to Animals Act. These animals are confined in very small cages, with inadequate lighting and very restricted space for exercise.' Despite this report, the circus travelled onwards.

In November '95 Akef's Circus entered Mozambique with three tigers, seven lions, two horses, ten dogs and a python, setting up for a one month stay. Owner Akef left, three months later, reportedly to make arrangements for his animals and staff to travel to Egypt. But he abandoned all nine members of staff and all his animals. He completely disappeared. All too soon the money to feed the animals and workers ran out. Every animal was left to starve to death.

The plight of the circus animals came to the attention of Elena Son, a local businesswoman. When she saw the state the animals were in, she immediately arranged for veterinary attention. Every animal was extremely malnourished, extremely thin and by now some were seriously ill. The tiger's eyes were sunken and he had discharge running from his nose and his eyes.

Elena remembers : 'When I first saw the animals, I was shocked. They were excessively thin and lethargic. One tiger, two-year-old Raj (who shortly died from chronic pneumonia) was in an appalling condition and the horses were skeletal. We thought we would feed them until someone came for them, but as time passed, we realised they all had been left there, in their cages, to die.'

© Jan Creamer / Animal Defenders

HANEN : *abandoned in her cage.*
But Animal Defenders were on their way!

Elena set to work contacting various international animal welfare organisations. Desperate pleas for help reached London. Tusk Force and Animal Defenders immediately sent cash to feed the animals and set about formulating a plan. While Tusk Force co-ordinated the London end and arranged for confiscation orders to be issued for the animals, Animal Defenders undertook to organise and finance the rescue.

After many telephone calls back and forth, a CITES confiscation order was issued, ordering circus owner Akef to make arrangements for his staff and his animals before the deadline date, or his animals would be seized.

The day before the deadline, the Animal Defenders' Rescue Team arrived in Mozambique. They learnt that despite Akef's promises that he would return to solve the problem, there was no sign of him. The animals, therefore, could be handed into the care of the Animal Defenders, on the understanding they would be taken to a place of safety.

Immediately, the Rescue Team set off for the circus site. There they found six remaining circus workers. (Three others had been arrested for alleged theft and were in prison).

One of the rescuers remembers : 'It was our first sight of the animals. We have seen animals in circuses all over the world and conditions are universally bad – elephants chained up, lions and tigers in small cages on the backs of lorries, dull-eyed husks of once magnificent creatures, reduced to pacing their way through the monotonous hours of caged life in hopeless despair. But the conditions these animals were in, were *particularly* distressing.'

'The cages were in the region of 6ft. long and 3ft. wide – a lion or tiger could barely stretch out. It is truly horrific that most of these animals had been kept in these cages for several years – some more than ten years. Meals and nights were spent in their travelling container.'

© Animal Defenders

'Our first sight of Hanen - pacing up and down in her cage.'

'The three horses – a Palomino stallion and a mare who had recently given birth to a foal – were in a particularly bad state and pitifully thin.'

'In two other cages were the dogs, extremely friendly and despite everything, reasonably healthy. They had lived in cages, only being let out for runs around the circus compound.'

MURPHY *: In a cage on the end of a solid metal container was this enormous African Python, curled up beneath her light, asleep.*
Murphy was the only one of four pythons who had made it to Mozambique alive.

© Jan Creamer / Animal Defenders

There had previously been threats to kill the animals and anyone who tried to help them, so the Rescue Team were grateful to find themselves surrounded by guards. They all decided that the safest method of transporting the animals was in the container where they had lived all their lives. Local transport firms and construction companies kindly arranged free use of a specialised crane and lorry.

Next, negotiations had to be undertaken with a local nightclub owner. He had suddenly arrived on the scene, announcing he was the official sponsor of Akef's Circus and since Akef owed him money for rent of land etc., he was planning to confiscate all of Akef's property for non-payment of debts. And this included all the animals!

Finally, the Rescue Team had the go-ahead to move the animals. First to go were the three horses – Butterscotch, Amber and Tigga. They had been kept in a tiny enclosure for months and therefore moved stiffly and nervously. The mare's bones protruded horribly through her skin as she walked.

'But all three horses were to begin a new life,' says Animal Defenders. 'At their new Sanctuary at The Sentro Hipico Riding School, they were led into their new, clean stables. They tucked into food and rested on fresh hay. At least three of the animals were safe.'

‘After more paperwork, it was the turn of the six lions – Akef (the owner had named the largest, most impressive looking lion after himself!) Hod, Navine, Darek, Faruuk, Saeed and the three tigers – Fazir, Nassell and Hanen. After living in cages and tiny outdoor runs for between six and thirteen years, they were being loaded for the most important journey of their lives. Our emotions were running high.’

‘Murphy, the huge python, was placed in a box for transportation, followed by the seven dogs, all loaded into their travelling cage.’

‘The moment had arrived! The huge transporter carrying all the animals, with a convoy of Rescue Team support vehicles, rumbled out of the circus site.’

‘We had been warned,’ remembers one of the Team, ‘to leave the country immediately. Akef’s family had apparently threatened to return to stop the rescue.’

When the convoy arrived at the border post, there was great excitement. Even the South African Immigration Officer stamped their passports and then offered to home all of the dogs! With a huge sense of relief they headed on into South Africa and when they halted for watering, the dogs went into the care of South Africa’s local SPCA, where all found new, loving, local homes.

Murphy was next. He went off to a wonderful retirement at the Fitzsimmons Snake Park.

Then, back on the road with the lions and tigers to continue their epic journey to freedom. By dusk, they had reached the Hoedspruit Research Centre for Endangered Species where there was an excited welcoming committee.

‘For a week we had seen these magnificent animals confined in their tiny runs and now we were literally carrying them to freedom,’ said one of the rescuers. ‘Every waking hour had centred on rescuing them, often not knowing if our attempts would be thwarted. Now they were safe at last. The Big Cats went through a battery of health checks and then it was time to search out a new permanent secure home. They were tame, dependent on people, so they could not be released back into the wild.'

It was during this process that it was realised that the lion Akef, with a huge black mane running past his shoulders, bore all of the distinct characteristics of the extinct-listed Barbary. Once common throughout North Africa, the last wild Barbary was killed by a hunter in 1921. The rescued lionesses too, had certain Barbary characteristics.

It was therefore decided that Akef and the lionesses should stay at the Centre for Endangered Species, with a view to establishing a breeding group from which the Barbary could eventually be returned to the wild.

A special thirteen-hectare enclosure of wild bush, with a waterhole, was selected for the lions into which they were released.

'As the Animal Defenders team drove to the enclosure, we were greeted with a wonderful sight – one of the lionesses stretched across a dirt road, basking in the sun. Memories flooded back of her in her tiny cage, pacing up and down, digging at the ground, desperate to escape. Akef strode out, proudly surveying his surroundings.'

'If these lions breed, they will eventually form a pride in this huge enclosure. From this, it is hoped that this magnificent breed of Barbary Lions can be reintroduced into the wild.'

The three other male lions were moved to the Moholoholo Wildlife Centre – a wonderful Sanctuary for unwanted and injured animals. 'These lions took their first steps into the real world of trees and natural vegetation. At first exploring, then settling down, concealed in the dappled shade of a bush.'

It was not possible to return the tigers to their wild or semi-wild homes in India, due to their unknown background. So the Animal Defenders chose the Pumula Game Lodge in KwaZulu Natal, where the tiger enclosures would be built to Animal Defenders' specifications. They would be happy at last. They now live in a very large enclosure of natural bush with a swimming pool, trees and a cave.

© Jan Creamer / Animal Defenders

Together and free. Akef, the Barbary Lion with one of the two Barbary Lionesses, enjoying their new home in South Africa.

'Our rescue of the Akef Egyptian Circus animals raises many important issues,' says Animal Defenders. 'Not just the way animals are treated but the fact that circuses can have such rare animals. The lions – the three survivors of an extinct-listed sub-species – got very close to starving to death in tiny cages in Mozambique. No-one would have realised that the Barbary Lion had slipped through mankind's fingers again.'

'With all that we know of the needs of both wild and domestic animals, how can we allow them to be shunted around the country in travelling menageries? It is little wonder that some of these animals go out of their minds. Lions, tigers, elephants and bears continue to live in chains or tiny cages on the backs of lorries.'

'Tradition cannot justify such misery. Surely people can enjoy themselves at a non-animal circus with exciting human acts instead?'

Happy at last in the Sanctary of the Pumula Game Lodge.

CURIOSITY & PATIENCE
AND TWENTY-THREE OTHER LITTLE DONKEYS

How the LORD WHISKY ANIMAL SANCTUARY saved twenty-five little Donkeys, stranded at Dover docks, on their way to Belgium meat traders.

The Lord Whisky Sanctuary Fund, a sanctuary for rescued animals and abandoned pets, was named after a little Jack Russell, Lord Whisky, who lived with Margaret Todd and her family for most of his twenty-two years. His history and character inspired Margaret to set up the sanctuary which, after fifteen years, continues to grow and grow.

Through their gates they have welcomed Mrs. Tinkywink, a snorty black pig with lots of character who will do anything for a banana! Sooty, a garden sheep who walks on a lead and loves cream cracker biscuits! Amy, a pretty grey mare ..... Not to forget the cheery chipmunks, marmoset monkeys, the elderly horses, rabbits, goats, cats, dogs, gerbils and a cockerel!

Here too, you will find two very special Donkeys – Patience and Curiosity, grazing in their paddock, with quite a story to tell.

Several years ago, a call came through to the sanctuary from a member of the public. Twenty-five little donkeys, having travelled all the way from Ireland, had arrived at Dover Docks. They were due to travel across to Belgium ...... for meat. But at the docks, they were turned away. A vet had pronounced them sick, so they were being sent to a local dealer just outside Canterbury to recover before the journey.

Margaret and her helpers set off for the docks immediately. Once there, she was told there was a lorry on the way from Ireland to take all the donkeys back home. 'Glancing at these poor, sad creatures, some of whom were unable to stand, we knew we *had* to help them,' she said.

The vet, having agreed that the majority were in no fit state to travel, provided emergency treatment and their temporary carers were advised they should not be moved. So Margaret and her team returned home and went into action.

'The next few days,' remembers Margaret, 'were quite amazing. We could not risk any infection being passed on to our large animals at the sanctuary, nor could we put these little donkeys anywhere they would be in close contact with other equines. But our Guardian Angel showed herself and we were offered two lovely paddocks, one to rent and one to use free of charge. A donkey shelter was quickly erected, transport was arranged and hay and straw ordered.'

But then things started to go wrong. The donkeys 'disappeared' and the horse-box was involved in an accident. But after a very worrying two days, the donkeys were found and a price agreed to purchase all twenty-five of them. Only then, could Margaret breathe a huge sigh of relief.

'I must admit, it was a lovely sight to see those twenty-five donkeys safe in their very nice accommodation in the Kent countryside. The time had now come for us to get to know them all individually.'

© Lord Whisky

The little Donkeys gather together safely in the paddock.

'Noddy, Big Ears, Joseph and Bourneville were looked after by people who became so involved and fond of them, they found they could not part with them!'

'Dosey went to live with three of our goats in a beautiful area where she can investigate lots of interesting paddocks and Patrick moved nearby to keep a horse company and can often be seen at our events.'

'Dandy and Daisy left for a home full of love. Little did we, or their new owners, realise they were getting three donkeys for the price of two! It soon became obvious Daisy *was* in foal and on a beautiful, sunny morning in April, her chocolate-brown filly, with a scattering of white hairs on her forehead and a white nose, was born directly beneath a blossoming cherry tree. Cherry blossom petals drifted down gently and decorated the foal's wet, curly coat – what else could they call her but Blossom?'

'Misty, a very shy little girl who needed to be on her own, continues to make good progress. Smokey, Tiny, Cindy and Lucy found homes together. Emma and Mary, two very sweet ones found a lovely home with walks in the woods, as did Bill, Ben, Coco and Minty.'

'Next to go were Moses and Amos who joined up with Patsy, a donkey who recently lost her companion – now all three are fussed over and spoilt. Choc Drop stole the hearts of the owners of the paddock and so will be staying there and Misty No. 2, also in foal, became a wonderful mum to little Dawn.'

'We wanted to choose homes where all the donkeys would get individual attention, not to be one of many animals; homes where there are people about, where people like to keep an eye on them. People who have become our friends and will keep in touch and people who care and are close to their animals.'

'We hope the future holds for them everything we would wish. We have tried our hardest to put right for them what should never have happened. They all have a return ticket and a little bit of their magic and faith will stay with us.'

In fact two of the donkeys never ever left. Margaret and her family found quite simply that they could not part with them all. So Curiosity and Patience are still at the sanctuary!

© Lord Whisky

PATIENCE : *Close to death.*

© Lord Whisky

PATIENCE : *Six months later, happy and content with Margaret's son.*

© Lord Whisky

CURIOSITY & PATIENCE
'Welcome to our Sanctuary!'

'When the donkeys arrived,' says Margaret, 'one was still very sick and together with a delightful little donkey called Curiosity, we willed her to live. We called her Patience, as each night, when she was put to bed, we asked her to be patient with us as life *would* be worth living if she hung on. She had a long way to go but Curiosity kept an eye on her for us.'

Patience and Curiosity are now the little stars of the sanctuary, absolutely adoring the attention from the many visitors to the tea rooms and museum. When Christmas came around, they were the first to attend the blessing service.

'Curiosity and Patience, along with several of our other animals, were blessed by the Vicar of Lyminge. It was wonderful to see them looking so well on what was really a most happy occasion for us. Patience had been really ill but she had become fit and well and made the congregation laugh by pulling faces during the singing. We weren't sure if that meant she liked the noise or not!'

Probably, she was just very happy to be there! All the twenty-five donkeys are now happy – with their foals they number twenty-eight, with Poppy being the last foal to be born. Her owner comments : 'Poppy can be a right little madam but you can't help loving her. She is very good at putting on a show for visitors and never ceases to be able to make us laugh each day.'

To the best of the sanctuary's knowledge, no donkeys have passed through the docks at Dover since the rescue.

'We thank everyone,' says Margaret, 'who has given them a loving home and made the donkey dream come true.'

DAISY DUMPLING
THE RABBIT

COTTONTAILS RESCUE's Rabbit with character.

'Cottontails Rescue was born in 1993,' says founder Mairwen Abbott. 'We take in approximately nine hundred unwanted rabbits a year – the highest number for any shelter throughout the whole of the UK. Rabbits come in from far and wide, chiefly because we have the expertise and ability to cope with large numbers. During the winter months, we often have to house one-hundred-and-forty rabbits, not including a waiting list of up to seventy rabbits needing a place at the shelter! It would be impossible to run Cottontails without the help and support of our regular volunteers. Their friendship and loyalty is much appreciated. With their help Daisy Dumpling proved to be one of our success stories.'

'Daisy Dumpling's mother arrived at Cottontails heavily pregnant and gave birth to four babies. Daisy growled at me even before his eyes were open so I had no doubts early on that here was a rabbit with character! Sadly, by the time he was four-weeks-old, it was apparent that he had severe health problems.'

'But by the time he was five-months-old, Daisy had survived two successful operations to remove his front teeth. Although he is tiny, he is a happy little bunny with a good quality of life. He was initially paired up with a blind baby called Ivy to keep him company but when Ivy died, Daisy was matched with two very docile guinea pigs.'

'It is working well because Daisy Dumpling can't bite them (no front teeth), he can't mount them (back leg problems) nor can he eat their greens.'

'A Perfect Match!!'

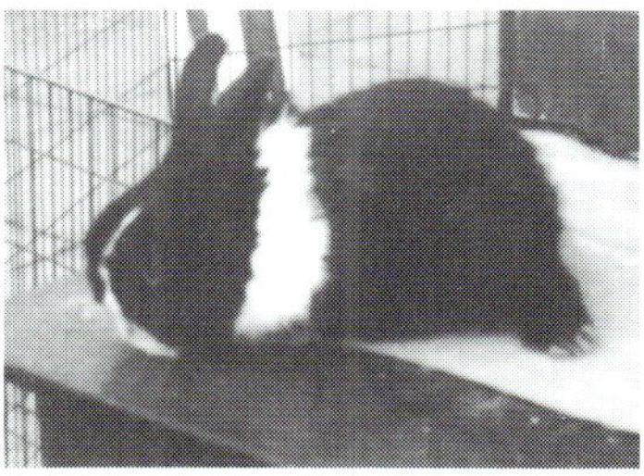

DAISY DUMPLING

BELLA
THE PORCUPINE

Helped, together with two porcupine friends,
by STICHTING AAP, who provide a Sanctuary
for the huge overflow of exotic animals
smuggled through The Netherlands.

Over one hundred species of exotic animals can be found at the Stichting AAP facilities in Almere in The Netherlands. Many years ago severe rules were imposed in the country regarding the import and export of all exotic animals. Exotic species smuggled into the harbours and airports, along with animals not native to The Netherlands, would all be confiscated. With the excessive amount of animals in animal sanctuaries, Stichting AAP stepped in and with such an overflow of exotics their new facility was opened.

Apes, monkeys, reptiles and small mammals have all found refuge here. They all have a sad history. Many are abandoned, abused or retrieved by custom officials. Care is not limited to illegal trafficking but also offered to exotics that have been sold as 'house pets'. People find baby monkeys lovable and sweet-tempered but when the animal matures, the infant-like behaviour changes and natural instincts manifest. Monkeys and apes will begin to exhibit certain behaviours indicative to group living. As a result, the animals get neglected and mistreated.

When animals arrive at Stichting AAP, they are often both physically and mentally disturbed. After a quarantine period, the animal will be placed with members of it's own species – a more natural atmosphere, providing a better and safer environment. Ideally, the foundation would prefer to reintroduce all the species back into the wild but in most cases they are too 'humanized' to ensure survival. So the aim here is to form a group of animals that can be sent to a nature reserve in their native land.

For many exotic animals, permanent injuries and loss of natural behaviours make relocation possibilities impossible. In this event, Stichting AAP becomes their permanent, caring home.

BINGO : *The Monkey.*
His 1" thick metal collar was so tight around his neck, it was surprising he had not choked to death. Now, in the safe hands of Stichting AAP, he is happy again.

Bella, and her two friends Istrice and Useppe, are three special porcupines. They came from a small animal park in East Germany that was formed over an old graveyard. After the reunification of East and West Germany, the land which included this animal park, was returned to it's previous owner. All the animals were destined to certain euthanasia due to the complete overflow of animals in the zoos. Few animal sanctuaries were able to take them in.

Stichting AAP, was the sanctuary that came to the rescue just in time. Along with the three porcupines, came an Indonesian Crowned Macaque, a red Coati Mundi bear – in poor health with a pelt full of holes and a Kinkajoe, who had spent his life in the office at the animal park, inside a cage the size of a kitchen cupboard.

'Many contributors to our Foundation,' said a Stichting AAP helper, 'made donations towards the travel arrangements so that Bella and the other porcupines could be rescued and transported to their new home. On arrival, they appeared to be most content in their spacious cage with one another.'

'Bella may seem rather passive sometimes but when threatened, her prickly rolled up ball, with a back resembling a coarse hairbrush with 30cm. long quills, can quickly stretch and separate. Beware! Bella can quickly become an 80cm. enemy! Anyone who challenges her could soon have serious injuries. In addition, she can produce a loud rattling sound from the quills on her tail and these defence mechanisms are able to ward off many a predator.'

The porcupines soon needed a lot more space. So Stichting AAP were lucky enough to find a very good place for Bella and her two porcupine companions. They moved to become residents in a zoo in Amsterdam. All three now have all the room and space they could wish for and much fun is had by all in tunnelling and tossing the sand around in their new enclosure. There couldn't be a happier trio of porcupines!

Porcupines are prevalent around Asia, Africa and Southern Europe. Now Stichting AAP are proud to say that they are present in The Netherlands!

© Marisa Bellis / Stichting AAP

BELLA : *The Porcupine saved from euthanasia.*

'At Stichting AAP, we care for a diverse variety of birds including Red Macaws, Green Amazon and African Grey Parrots, and Cockatoos. Although the species are different in nature, they all share one element in common – all have been abandoned. We acknowledge that birds are social creatures, happiest with members of their own species. When they are bought as house pets, their whistles, screeching and talking, due to boredom, reach an intolerable level.'

'Desperate owners resort to locking the birds in dark places, such as garages.'

'Lorre, a Green Amazon parrot created a constant noise. A desperate husband told his wife that either Lorre leaves or he would. Needless to say, Lorre ended up with us!'

© Marisa Bellis / Stichting AAP

'Mynah birds are one of the most vocal of all birds, able to produce an unbelievable sound.'

'Noach's voice was so penetrating that his owners handed him over to us and now Noach lives with other Mynah birds where he is able to screech to his heart's content!!'

'From time to time we are able to place these birds in nature reserves but for many Stichting AAP will be their permanent home.'

'Throughout the year, the upkeep of one exotic animal alone can exceed thousands of guilders,' says Daphne Lavy, the animal adoption co-ordinator, 'so many exotic animals and birds here can be supported through our adoption schemes. They have the right to a decent life so our supporters have a part in providing that life. Our goal, as always, is to turn sad histories into new beginnings.'

SIMPLY SIMON
THE LURCHER PUP

LAST CHANCE ANIMAL RESCUE was there when he needed help. Now he gives so much back in return.

Sylvia Wragg, who runs Last Chance Animal Rescue in Kent, is the first to admit that owning, running and living at an animal rescue home is no bed of roses. But the story of 'Simply Simon' makes it all so worthwhile.

'A few years ago,' remembers Sylvia, 'a very distressed lady telephoned me regarding a horse fair. Apparently, her home backed onto the land where the fair was due to be held and each year she found herself witnessing the poor dogs that were taken to be sold and subsequently abandoned, if no buyer was found. She told me of puppies left in crates, Lurchers running around starving and beaten, with abused dogs of every shape and size. She begged me to visit the fair and do whatever I could.'

'Why people think that my heart is stronger than theirs when experiencing such sadness, I shall never know, but the thought of these poor animals haunted me. I found myself driving the one-hundred-and-fifty miles to the sales!'

Sylvia found that the horse fair was just as the lady had described. Puppies and adult dogs were being dragged around on ropes or in crates, offered to anyone who looked even half-way interested. There were dogs that cringed and cowered and dogs with eyes like zombies. 'I tried to look away. I couldn't possibly rescue them all,' says Sylvia. 'Heaven knows, I wanted to!'

At the end of the day, a tiny Jack Russell and a brindle Lurcher were left tied to the railings. A rough coated Lurcher pup sat abandoned in a crate. 'I took all three to my van and then tried to catch others which were running free. It was impossible; they were so frightened, they ran wildly on.'

On arriving home, the tiny Jack Russell, now named Sugar Lump, was checked over by the vet. She had probably been kicked as the muscles in her stomach had ruptured. The Lurcher, named Good Grief Reif! was fortunately just simply underweight. Both dogs were soon to find good, new homes.

'But the little pup,' says Sylvia, 'was worm-ridden, covered in mange and absolutely terrified. He needed special nursing, during which time he burrowed gently into my heart.'

'I named him Simon, just simply Simon, after the best human friend I have ever had.'

'The more Simon crept under my skin, the longer his legs grew. By the time he was six-months-old, he and I were inseparable.'

About this time, Sylvia was due to do a three-hundred-mile sponsored walk. When the time came, she could not bear to leave her new friend. So while she did the walking, her daughter Leah drove a little cart pulled by two miniature Shetland ponies. Simon sat happily aboard, enjoying the view! They managed to raise £14,000 for animals needing help.

The following year, the bond between Sylvia and Simon was even greater. Simon joined her, with their Shetland pack pony, on a walk that carried them over 1,000 miles from Scotland, down through Wales, to Lands End. They raised a massive £21,000 for needy animals.

'Because of Simon's devotion and friendship, I have been able to carry on from one day to another, rescuing hundreds of animals. We set up our own veterinary clinic at Last Chance, to help those who cannot afford to help themselves and of course, there is that endless stream of unwanted and abused animals.'

'During the nights, when I lie awake worrying about the next day, it is Simon's whiskery face that is the first to push under my hand and comfort my troubled mind.'

'Life is a circle. I rescued Simon and now, most days, Simon rescues me!'

SIMON

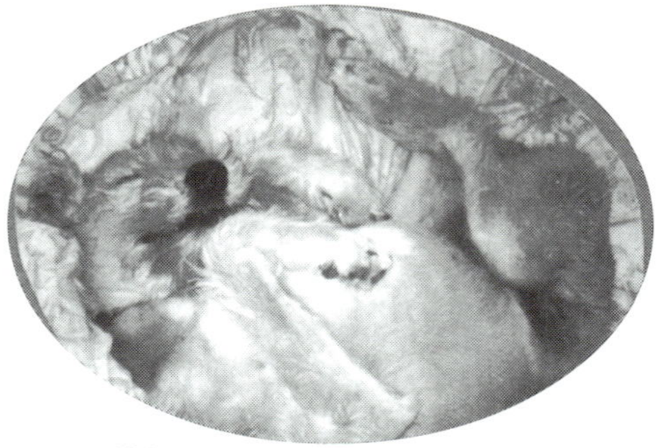

NIDGE, SYLVIA & SIMON : *Snatch a quick rest during their sponsored walk.*

Well Done Chaps! We've finally made it! 1040 miles.

ACKNOWLEDGEMENTS

The Author would like to thank the following individuals and the many Animal Sanctuaries listed below, for their help in the compilation of this book.

ANIMAL DEFENDERS / NATIONAL ANTI-VIVISECTION SOCIETY (NAVS) London W12 9PE	Gill Freshwater
CHEETALERT! The Lifeline Survival Trust, Minehead, Somerset TA24 8SB	David Nelhams
COTTONTAILS SANCTUARY Fishponds, Bristol BS16 5AH	Mairwen Abbott
FARM ANIMAL WELFARE NETWORK (FAWN) / CHICKENS' LIB Holmfirth, Huddersfield HD7 1QY	Clare Druce
FRIENDS OF THE ANIMALS Bearwood, Warley, West Midlands B66 4EX	Helen Sinclair
GREEK ANIMAL RESCUE (GAR) Hendon, London NW4 1PT	Vesna Jones
HEDGEHOG CARE Authorpe, Louth, Lincs. LN11 8PF	Elaine Drewery
HILLSIDE ANIMAL SANCTUARY Frettenham, Norfolk NR12 7LT	Wendy Valentine
INTERNATIONAL OTTER SURVIVAL FUND (IOSF) Broadford, Isle of Skye IV49 9AQ	Grace Yoxon
LAST CHANCE ANIMAL RESCUE Edenbridge, Kent TN8 5NH	Sylvia Wragg
LIFE FOR CATS Barnet, Herts. EN5 5TS	Peri Ann Sinclair
LORD WHISKY SANCTUARY FUND Stelling Minnis, Kent CT4 6AN	Margaret Todd MBE
MEDASSET London W1Y 7DF	Lily Venizelos
NATIONAL ANIMAL WELFARE TRUST (NAWT) Watford, Herts. WD2 8HQ	Franca Cappuccini

NATIONAL CANINE DEFENCE LEAGUE (NCDL) London EC1V 7LT	Clarissa Baldwin
NOAH'S ARK – RAT RESCUE CENTRE Downend, Bristol BS16 6RX	Beverley Taylor
PROTECTION OF ANIMAL LIFE SOCIETY (PALS) Newton Tony, Salisbury SP4 OHW	Lorraine Drake
PINE RIDGE DOG SANCTUARY Ascot, Berkshire SL5 8RJ	Connie Cuff
REDWINGS HORSE SANCTUARY Frettenham, Norfolk NR12 7RW	Anne Galloway
RHODES ANIMAL WELFARE SOCIETY (RAWS) Rhodes, Greece GR85100	Maggie Avgerinou
SOUTH DEVON ANIMAL SHELTER Ashburton, Devon TQ13 7BP	Joy Clift
STICHTING AAP Almere, The Netherlands	Petra Buckers
THE ELEPHANT SANCTUARY Hohenwald, Tennessee, USA	Carol Buckley
WILDLIFE AID Leatherhead, Surrey KT22 OAL	Susan Blissett
WINSLADE WILDLIFE SANCTUARY Nr.Buckland Brewer, Devon EX39 5NH	Tracey Berridge
WORLD SOCIETY FOR THE PROTECTION OF ANIMALS (WSPA) London SW8 1TJ	Jonathan Owen

Linda Thurlow
Marilyn Preston Evans
Sharon Callahan

Cover Design: HolRocks Books

Front Cover Illustration : Linda Thurlow

ALSO AVAILABLE FROM HOLROCKS BOOKS (B)
P.O.BOX 107, PAIGNTON, DEVON TQ4 7YR

CELEBRITY PETS

Claire McClennan

The Dogs and Cats who brighten the lives of some of Britain's popular personalities : Ernie Wise, Ken Dodd, June Whitfield, Anthea Turner, Terry Waite, Lorraine Kelly, Kriss Akabusi, and many more.

'A nice gentle read about famous personalities and their pets.'
Editor, Pet Dogs

'The book is lovely.'
Eastenders Actress, Wendy Richard

''Tis a great wee gift for animal loving pals.'
Actress, Pat Coombs

'A lovely book.'
Radio/TV Presenter, Valerie Singleton

£6.99 ISBN : 0-9526944-0-9

ANOTHER CHANCE

Claire McClennan

A COLLECTION OF TRUE DOG STORIES –
Abandoned, neglected, victims of cruelty, rescued by animal lovers and offered a new happy life.

A Lovely Gift Book – Many Photos – Detailed Artwork.

'Delightful! We are sure it will be very popular.'
NAWT

'We enjoyed the book very much.'
Friends of the Animals

'Heartening tales of Dogs. *Another Chance* will appeal to anyone who cares about animals.'
PDSA

£6.99 ISBN : 0-9526944-1-7